Walch Science Literacy Series
Earth Science

by Dr. Richard Ojakangas
Department of Geology
University of Minnesota, Duluth
Duluth, Minnesota 55812

illustrated by Carol Stutz

J. WESTON
WALCH
PUBLISHER
Portland, Maine

To the Student

How would you like to be one of the best thinkers in your school? Would you like to be able to put pieces of a problem together quickly and thoroughly in order to find a solution? Do you want to be able to spot flaws in weak arguments? Can you develop a strategy for setting up an experiment that will work to give you an answer to a problem?

Most students would like to be able to do all of these things. But some of them are not as successful as they could be because they have not developed the skills—the "habits of mind"— of really good thinkers. In this program you will learn some new thinking skills, and you will know when you are using them. You will become a more effective thinker and problem solver as you work through the science situations in this book.

Our best wishes for good thinking.

1 2 3 4 5 6 7 8 9 10
ISBN 0-8251-3309-2

J. Weston Walch, Publisher
P. O. Box 658 • Portland, Maine 04104-0658
Printed in the United States of America

CONTENTS

CONTENTS

Lesson 1

How Do We Know How Old Rocks Are?

Asteroid makes dinosaurs extinct

BERKELEY—Scientists at the University of California at Berkeley think they know why dinosaurs disappeared about 65 million years ago.

They have found evidence that an asteroid from outer space crashed into the earth. The crash probably sent clouds of dust and dirt into the air, where they lingered for months.

Scientists speculate that this pollution changed the earth's atmosphere enough to cause dinosaurs to become extinct.

How has the earth changed over time? Are the same processes that changed the earth in the past still changing the earth today? Clues to what happened on earth millions of years ago are found in rocks that have existed since those times. You could say that the present is the key to the past.

Sedimentary rock was once layers of mud or sand that were squeezed together under great pressure. Sometime you can see ripple marks or mud cracks in sedimentary rock. How did they get there? If you go to a beach, you might see ripple marks in the sand. These marks could be preserved if the sand turned into sandstone, one kind of sedimentary rock. The mud cracks you see in puddles could be preserved if the mud became shale, another kind of sedimentary rock.

Some layers of sedimentary rock contain **fossils,** the remains and traces of ancient life. If you see a dinosaur fossil in a rock, you know that a dinosaur once lived where the rock was found. But how many years ago was that?

Over 200 years ago, scientists started constructing the **geologic time scale** to show the order of events in the earth's history. By studying layers of sedimentary rock, they could determine a layer's relative age—that is whether one layer was older or younger than another layer. Usually, the oldest layers are at the bottom of a rock formation and the youngest are at the top.

Scientists can also tell the relative ages of **igneous rocks,** or rocks that formed from molten or melted material. Suppose a black igneous rock cuts across a pink igneous rock. Which rock would you guess is older? If you say the pink rock is older, you are correct. The pink rock had to be there before the black rock could cut across it. The rule of cross-cutting relationships tells you that an igneous rock is younger than the rock it cuts across.

Now you know how scientists find the relative ages of rocks. They have another way of finding out the numerical age, or how old the rocks actually are in years. Rocks are made up of atoms. Some rocks contain radioactive isotopes, forms of elements whose atoms change over time. The atoms of radioactive elements decay, or break apart, to form other elements. The length of time it takes for half the atoms of a radioactive element to decay is the **half-life** of the element.

Scientists determine the age of a rock by measuring the half-life of an element the rock contains. One such element is uranium 238. Scientists have learned that it takes 4.5 billion years for half of a given amount of uranium 238 to decay into lead 206. The half-life of uranium 238 is therefore 4.5 billion years. If a rock contains as much lead 206 as it does uranium 238, you know that half the uranium 238 must have changed into lead. Thus, the rock sample is 4.5 billion years old. This method of finding the numerical ages of rocks—based on radioactive decay—was developed in the middle of the 1900's. So, the numbers are rather new additions to the geologic time scale.

Now it's your turn to answer questions about the ages of rock.

1. Imagine that you are a geologist working in the Rocky Mountains. The rock layers shown below are exposed on a cliff. Rocks 1 and 2 are igneous, and rocks 3, 4, and 5 are sedimentary layers.

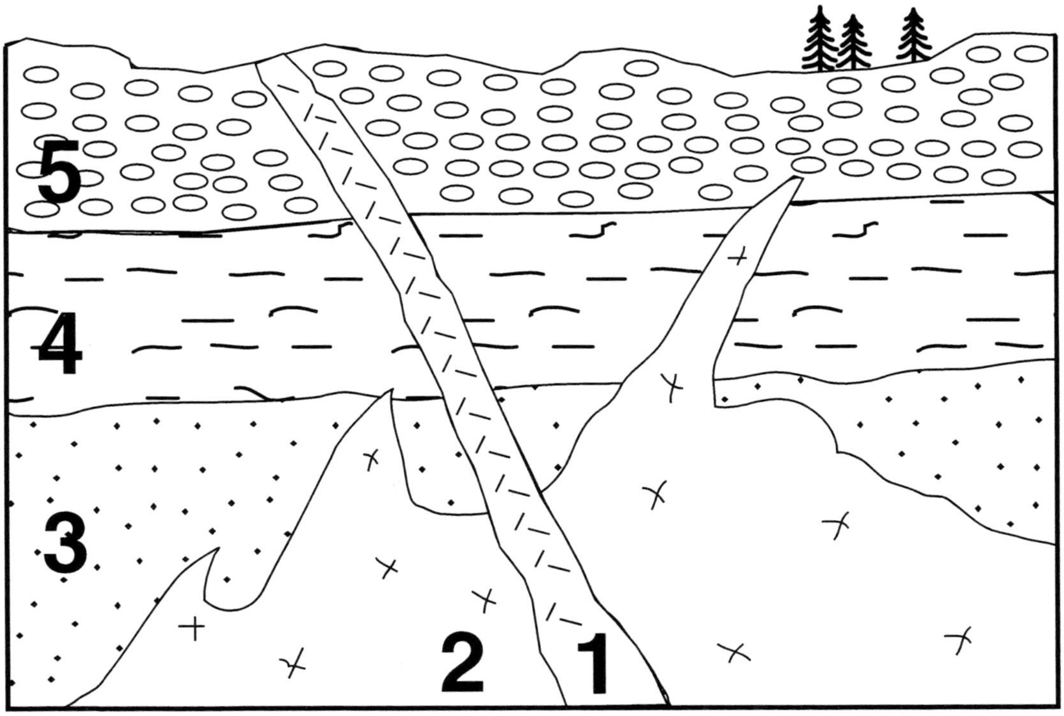

(a) Determine the relative ages of the five rocks in the sketch. List them in order of age, from oldest to youngest.

____

____

____

(b) If rock 2 is 60 million years old and rock 1 is 40 million years old, what can be said about the ages of rocks 3, 4, and 5?

____

____

____

Lesson 2

The Earth Underfoot

As you stand on the surface of the earth, you have no way of knowing for certain what is beneath your feet. Over many years, geologists have collected, studied, classified, and analyzed samples of rocks and minerals found in the earth. By now, they have a clear picture of the materials that make up the earth's rocky crust.

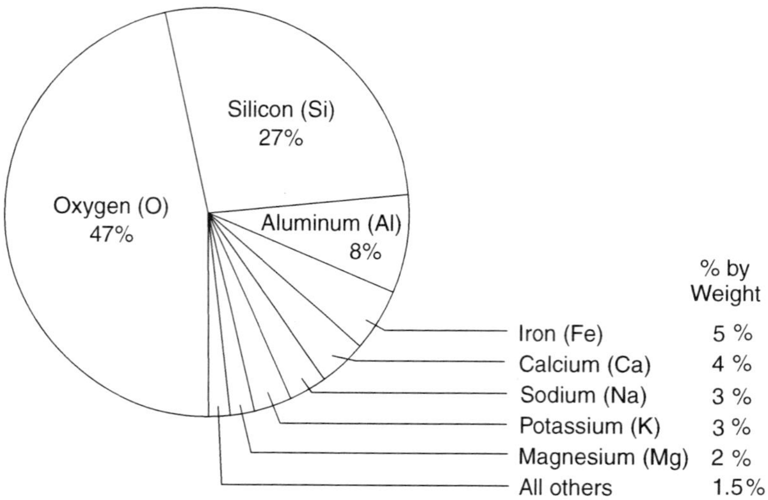

Percent by weight of common elements in the earth's crust

Rocks are made of **minerals,** and minerals are made of **elements.** There are 92 naturally occurring elements (and another 16 that have been made in nuclear reactors). You probably recognize elements like gold, silver, aluminum, hydrogen, oxygen, copper, and many others. What elements do you think are most common in the earth's crust?

Which element makes up nearly half of the rocky crust by weight? Does it surprise you that it's oxygen? You may be used to thinking of oxygen as an element in air, not in rock. As you can also see from the graph, 27 percent of the earth's rocky crust by weight is silicon and 8 percent is aluminum. In fact, 98.5 percent of the crust is made up of only 8 elements.

These elements do not appear in their pure forms in the earth's crust, but they are combined to form the minerals of which rocks are made. The elements combine at the atomic level. The most basic mineral building block in the earth's crust is a combination of oxygen and silicon, called the **silicon-oxygen tetrahedron.** It is made of one silicon atom surrounded by four oxygen atoms, and its atomic formula is written SiO_4.

Tetrahedra sometimes "share" oxygen atoms. That is, one oxygen atom may be part of two different tetrahedra. If all oxygens are shared, the formula is SiO_2. This is the formula for quartz, a common mineral.

In all, there are about 2500 minerals, but only 10 of them make up almost all of the earth's crust. And 9 out of the 10 are silicates, minerals containing silicon and oxygen. This makes the silicate minerals the main "rock formers."

In addition to silicates, there are five other major mineral groups: oxides, sulfides, halides, carbonates, and sulfates. Each group gets its name from the elements that form the minerals. For example, oxides form when an element combines with oxygen. Oxides include corundum, hematite, and magnetite. Sulfides are minerals that contain the element sulfur. Sulfides include galena and pyrite. Halides form when certain elements combine with chlorine, iodine, bromine, or fluorine. Halides include halite and fluorite. Carbonates form when an element combines with a carbon atom that is surrounded by three oxygen atoms. Carbonates include malachite, azurite, and aragonite. Sulfates form when an element combines with a sulfur atom surrounded by four oxygen atoms. Sulfates include gypsum and barite.

Now you may have a better idea of what makes up the rocks beneath your feet. They are made of minerals, which in turn consist of elements that are bonded together at the atomic level. In some rocks, such as granite, you can actually see the minerals. They are the many small particles of different colors and textures that are pressed together in the rock. In some rocks, the particles are too small to see. But you can find out what a mystery rock contains by performing tests to determine its hardness, color, and other characteristics.

1. When you first think of minerals, you may think of minerals that are ores, such as iron ore, which have economic value. List the common ores you can think of. Look in a rock-and-mineral reference book to find more.

____

____

____

____

2. Following is a chart of the 10 most common minerals in the earth's crust. Choose one of the minerals listed and find out about its structure. In what kinds of rock is it found? What can you say about where that kind of rock is found? In addition to its color, what characteristics does it lend to a rock? In your notebook, write a few sentences describing your mineral and give its atomic formula.

The Most Common Rock-Forming Minerals		
Name	**Elements Present**	**Common Color**
muscovite mica	H, K, Al, Si, O	clear
biotite mica	H, K, Mg, Fe, Al, Si, O	brown or black
orthoclase feldspar	K, Al, Si, O	pink
plagioclase feldspar	Ca, Na, Al, Si, O	white or gray
amphiboles	Ca, Na, Mg, Fe, Al, Si, O	black
pyroxenes	Ca, Mg, Fe, Al, Si, O	green or black
olivine	Mg, Fe, Si, O	green
quartz	Si, O	white or any color
clays	H, Al, Si, O (and some Ca, Mg, Fe, Al, Na, K)	gray or brown (white if pure)
calcite	Ca, C, O	white

Lesson 3

Rocks Are All Around You

Igneous, Sedimentary, and Metamorphic Rocks

Rocks are easy to find. Pebbles and small stones one or two inches in diameter are nearly everywhere. Good places to look for them are along the shoulders or ditches of roads, along stream valleys, or even in the schoolyard.

What story could a rock tell if it could talk? Was it "born" in a volcano? in the ocean? You can tell its story by describing it. Then identify what kind of rock it is. That is what geologists do when they study rocks. Then they can interpret the rock's history.

There are three main groups of rocks, based on how they were formed. These groups are igneous, sedimentary, and metamorphic. **Igneous rocks** crystallize from molten rock, or **magma. Sedimentary rocks** are made of **sediment,** small pieces of other kinds of rocks. **Metamorphic rocks** are rocks that have been changed from other kinds of rock by pressure and temperature.

When you think of magma, do you think of a red-hot lava flow pouring out of a volcano? That does indeed happen. The magma then comes in contact with air or water at the earth's surface. This results in very small, microscopic crystals. Why? Because the magma cools fast and the atoms don't have time to join together to form large crystals. Some magma is thrown out of volcanoes during volcanic explosions and forms volcanic glass. The magma in this case cools so fast that there is not enough time for any crystals to form. Much magma never reaches the earth's surface. It cools very slowly deep in the ground and so the crystals grow larger. These crystals can be seen with the naked eye.

Sedimentary rocks form on the earth's surface. They are made of particles that have been compacted or cemented together. The particles in the sediment come from the breakdown of rocks that are exposed to air, water, and processes of erosion. However, some sediment is precipitated out of water, commonly in the oceans. Only sedimentary rocks contain fossils.

Metamorphic rocks are formed from any other rocks below the earth's surface, where pressures and temperatures are very high. Metamorphic rocks have changed under these conditions. Because pressures are commonly stronger in one direction, the minerals in metamorphic rocks are commonly rearranged into parallel layers or bands.

Sedimentary rocks cover three quarters of the earth's surface in a thin layer. Some volcanic rocks are also on the earth's surface. In some places, metamorphic and igneous rocks are exposed by the uplift of the land and the erosion that follows.

All rocks are related. The "family tree" of rocks is commonly shown as the **rock cycle.** This shows that rocks can change from one type to another depending on the environmental conditions. Rocks are stable only in the environments in which they form. Follow the arrows on the rock cycle to see how rocks can change.

Study the rock cycle diagram to answer the following questions. Write the answers in your notebook.

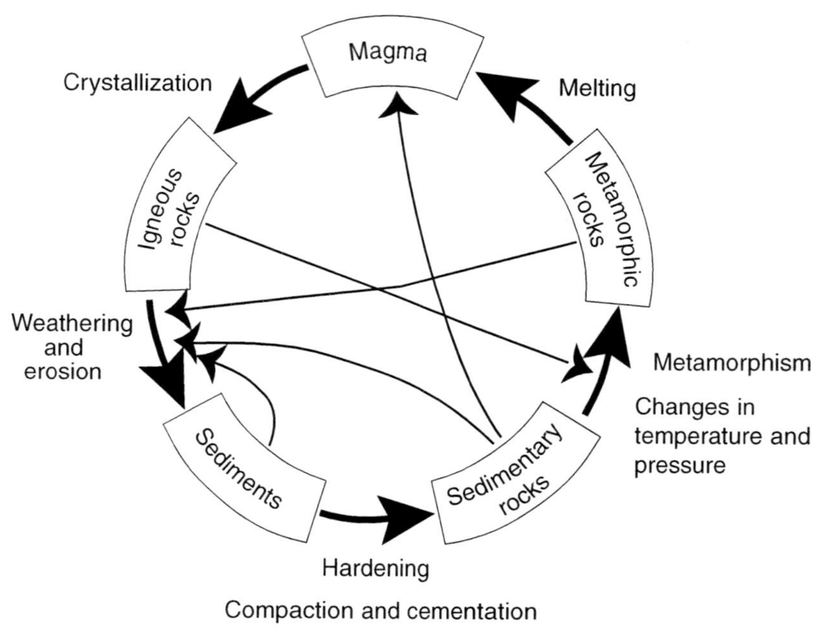

1. Does every igneous rock become a sediment? Explain.
2. If a rock traveled through the entire rock cycle and melted, would we know its history? Explain your answer.
3. Why do you think fossils are not found in meta morphic rocks?
4. What does the presence of small crystals in an igneous rock tell you about the rock's history?
5. What processes act on rocks to change them?

Lesson 4

Be a Detective—Look for Clues

The Present as a Key to the Past

Imagine that you are walking in a low muddy area after a flood. The mud is drying. You notice cracks in the mud. Some places are still under water and are not cracked. A clue! Could the cracks be the result of the mud shrinking as it dries? You think out loud, "So that is how **mud cracks** form. The water evaporates."

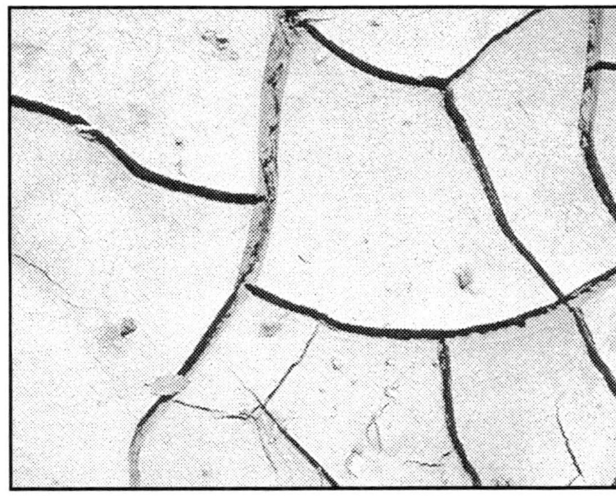

Now imagine that you are in the Wasatch Mountains of Utah, near Salt Lake City. The layers of sedimentary rock are vertical, standing on end. Were they deposited that way? That doesn't seem logical, does it? Other sediment layers you have seen were quite horizontal. Then you say, "Hey, I'm in a mountain range. The mountains must have been pushed up. The layers were tilted during mountain building." As you look more closely, you see a vertical layer with mud cracks. What is your conclusion? Is it possible that the rock you are looking at was once mud? A geologic map shows that the rocks are 600 million years old.

As you learned from the mud cracks, the present is the key to the past. This principle forms the basis for much work by geologists studying sedimentary rocks. They observe natural processes at work today. Then they use their observations to interpret the history of ancient sedimentary rocks. It is detective work.

James Hutton of Scotland first put this principle into print in the late 1700's. He is now known as the father of modern geology. He learned this principle of nature by observing and thinking.

Imagine that you are in a field of sand dunes in Idaho. You are climbing over big dunes, as shown here. You walk up the gentle sides of the dunes. You slide down the steep sides. You see how the sand is being blown over the top of the dune. You cut into one dune with your geology hammer. You see it is made of thin inclined sand layers, parallel to the dune surface. You look at the sand with a magnifying glass and see that many grains are rounded. They may have been rubbed together and polished by strong winds, their original sharp edges being rounded. Good observations and good clues.

Now imagine that you are a geologist in West Africa looking for oil. You are in the country of Mali. Out in the dry and wild countryside are some hills made of sandstone. You climb one of them. The sand grains are rounded. It all reminds you of something you've seen before. Yes—that's it! The sand dunes you once climbed in Idaho! But these rocks are 600 million years old. You conclude that quite likely a field of sand dunes was here 600 million years ago. You say to yourself, "The present really is the key to the past." You also see that the sand is in inclined layers, or cross-beds. You can use the **cross-bedding** observation to determine the wind direction (**paleocurrent** or old current) when the sand was deposited.

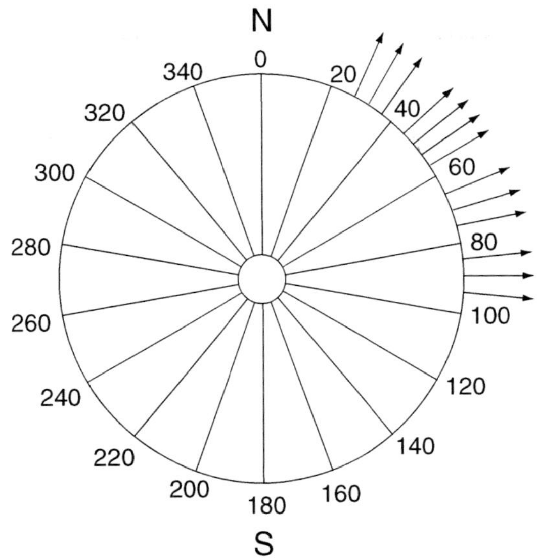

Plotting of paleowind direction on compass face

1. Express the principle "the present is the key to the past" in your own words.

 ____

 ____

2. Imagine that you have often observed small, shelled animals on the ocean beaches in Florida. Later, in Kentucky, you notice layers of coquina, a type of limestone, in a road cut. You observe that the coquina is made of fragments of shells. Use the principle "the present is the key to the past" to describe how the coquina limestone probably formed and what it can tell you about the geologic history of the area where you found the rock.

 ____

 ____

3. Do you think geologists who study volcanic rocks can use the principle "the present is the key to the past?" Why or why not?

 ____

 ____

Lesson 5

You Are Constantly Moving

You are constantly moving. Even when you are standing still, you are moving. "Of course," you may say, "the earth is rotating and it is moving around the sun." And you are right. But, even if the earth were standing still, you would be moving. Not much. But each year, North America moves westward about one inch farther away from Europe.

How did we come to know about the movements of continents? To answer that question, we have to go back in time.

The year was 1923. Alfred Wegener walked to the podium and began to address a group of important scientists. When he finished presenting his ideas, he was met with such comments as "Utter rot!" from the group's president, and "Ludicrous!" from a professor at the University of Chicago. Another geologist stated, "Wegener is blind to every fact and every argument."

What had upset the scientific community so much that Wegener was met with hostility by geologists everywhere he went? His speech had presented his idea of **continental drift.** He said that the earth's continents were not fixed in place. Instead, he said, they had, over millions of years, drifted around over the earth's surface. Further, he said, the continents had once been part of a single large landmass he called **Pangaea.** He said that the continents started drifting apart about 200 million years ago and that they are still drifting. Wegener presented evidence to support his theory. First, Wegener argued that the jigsaw-puzzle fit of the continents, especially South America and Africa, showed the results of the breakup of Pangaea. Second, he pointed out that explorers had found fossils and rock layers on the east coast of South America that were similar to those found on the west coast of Africa. Third, he described how other explorers had found rocks made of glacial sediments in lands near the equator.

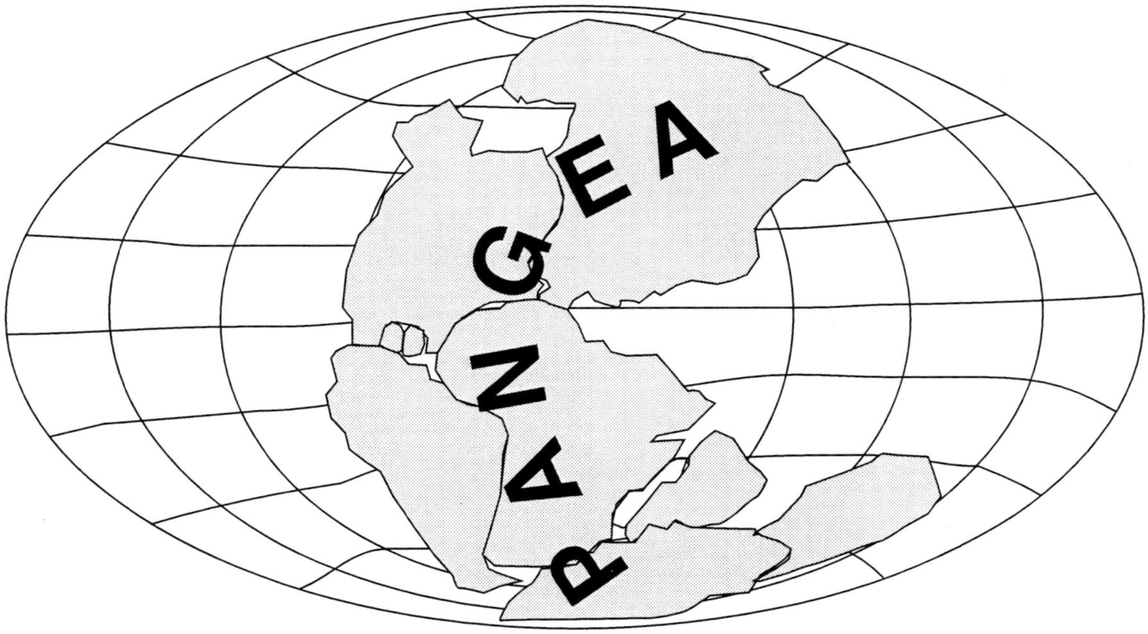

While Wegener's evidence was interesting, he could not explain *how* continents could move. When Wegener died on an expedition in the middle of Greenland's ice cap in 1930, his idea of continental drift nearly died with him. Few other scientists had accepted his revolutionary ideas.

Then, in the late 1950's and 1960's, the idea of moving continents was reborn. Why? New data had been collected, especially from the ocean floor. Rocks on the ocean floor contain iron. These rocks contain a record of the earth's magnetic field at the time that they were formed. So, scientists can tell where the magnetic north pole was for rocks from different continents. Shouldn't rocks of the same age from different continents indicate the same magnetic north pole? They should, but they don't. Therefore, the continents must have moved relative to each other. And, if North America is moved back toward Europe, the poles match up.

On the ocean floors, parallel bands of normally polarized rock alternate with bands of reversely polarized rock. The normal bands show that the magnetic north pole was about where it is today. The reverse bands show that the magnetic field was reversed. That is, a compass needle would have pointed south when those rocks were forming. And the bands on each side of the mid-ocean ridges are not only parallel to each other but also appear in matching pairs on both sides of the ridges. Age dating showed that new volcanic crust was being formed at the ridges. These are now called spread centers. This process is called **seafloor spreading.**

As the new information was analyzed, it fit the idea of the continents moving apart. This and other evidence led to **plate tectonic theory.** Plates are large sections of the earth's surface, about 100 km (60 miles) thick. *Tectonics* means large-scale movements. You can see the major plates on the map. Plates can move apart (spreading boundary), or toward each other with one plate *subducted*—pushed down—one under the other (colliding boundary), or sideways (strike-slip boundary).

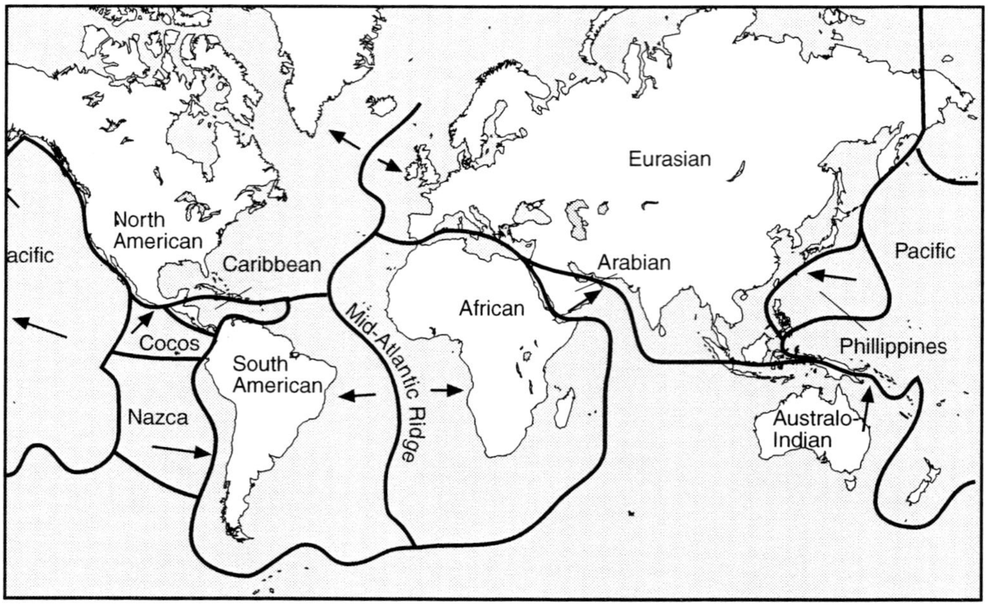

1. Describe what you think would happen to the earth if seafloor spreading continued at the spread center and there were no **subduction** of plates.

____

____

____

2. Compare the idea of **continental drift** to the theory of plate tectonics.

____

____

____

Lesson 6

When the Land Shakes

The house is shaking. Dishes are rattling in the cupboard. Books start falling off shelves. Your dog is whining, with her tail between her legs. You're having trouble keeping your balance. Oh, no! It's an earthquake! You grab the dog and race for the door to get outside and away from buildings. You don't want anything to fall on your head.

You could, indeed, experience an earthquake. Maybe you already have. It's most likely if you live in the western United States. And it is especially likely if you live in California or Alaska.

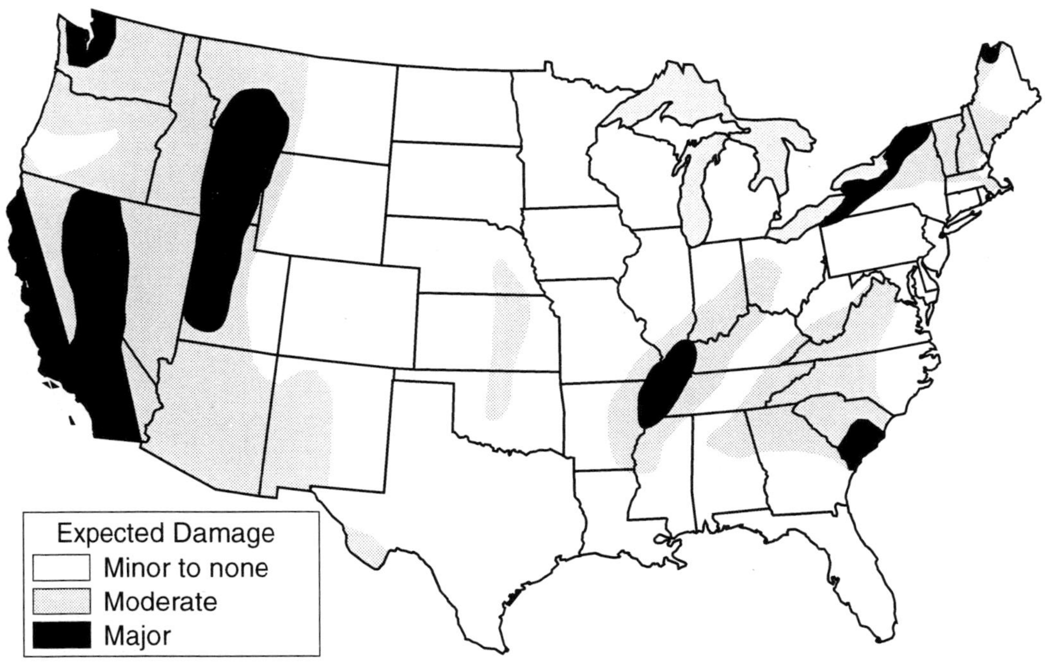

Map showing earthquake danger—not on map: Alaska, with damage rating similar to California's

What causes an earthquake? Many small earthquakes occur in magma chambers beneath volcanoes. They signal a possible eruption. But the big ones occur along faults. **Faults** are cracks in the earth's rocky crust, along which there is movement of the rocks. An earthquake occurs when rocks on opposite sides of a fault slide past each other. But why do they slide?

Pressures build up because of internal earth movements. Friction between the rocks on opposite sides of a fault makes the rocks stick and may prevent movement for a time. Rocks a short distance away from the fault may be slowly bending. Then, once the pressures become greater than the friction, there is sudden movement along the fault. This sudden movement causes shock waves that cause the ground and the underlying rock in that area to vibrate and shake. The bent rocks straighten out.

Geologists measure the energy released in an earthquake by the amount of ground motion. This motion is recorded on a **seismograph,** amplified, and printed out as a **seismogram.** The measured "wiggle," or shake, on the seismogram is then related to the **Richter scale.**

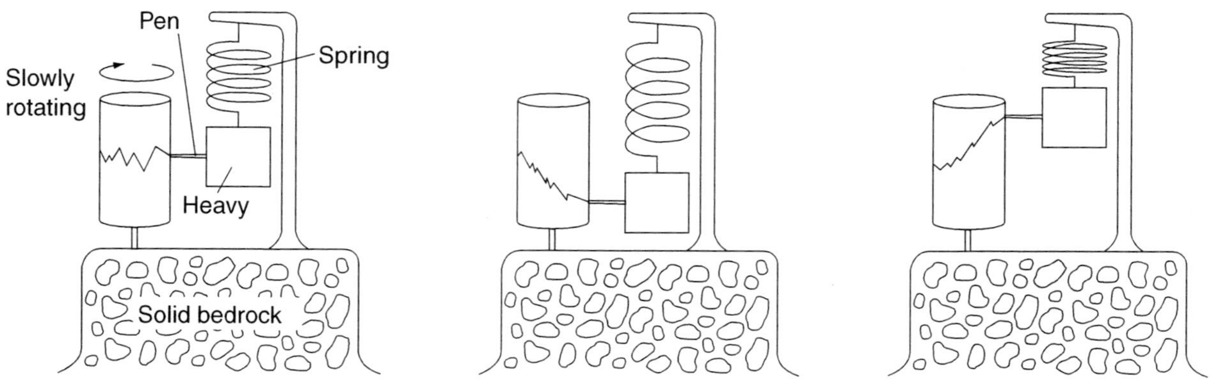

Seismograph measuring earthquakes

On the Richter scale, each number indicates a vibration 10 times greater than the previous whole number. Thus, an earthquake of 5.0 has 10 times the vibration of an earthquake of 4.0. And one of 6.0 has 10 times 10, or 100 times the shaking of a 4.0 earthquake. An earthquake of 6.0 is usually damaging. One of magnitude 7.0 can cause severe damage.

More importantly, the energy released during an earthquake is about 30 times greater for each whole number. Thus an earthquake of 6.0 releases 30 times 30, or 900 times as much energy as one of 4.0. The largest earthquake ever measured was 8.6 in southern Chile in 1960. An earthquake of about 2.0 to 2.5 can be felt by people on the ground. Many smaller ones are not even felt.

1. In what parts of the United States would you face the most danger from earthquakes?

____

____

____

____

____

____

2. Why do earthquakes occur?

____

____

____

____

3. How much more energy is released in an earthquake of 8.0 compared to one of 4.0?

____

____

4. What could you infer about the earth's crust in California from the fact that it has so many earthquakes?

____

____

____

____

____

Lesson 7

Ring of Fire

Imagine that you live on the southern coast of Iceland. The year is 1963. You wake up one morning in November and look outside at the ocean. The normally cold waters of the North Atlantic are boiling! Plumes of steam are rising from the surface of the ocean. What is happening? A volcano erupting on the ocean floor has heated the water to the boiling point. By the next evening, a ridge of land appears through the waves. A new island is being born.

Although you've probably never seen a volcanic island being formed, you may have seen photographs of volcanoes. When you think of volcanoes, you might think of a mountain with the shape of an upside-down ice-cream cone. But not all volcanoes form the same shape. Some volcano cones have steep sides; others have smooth domes. The way a volcano erupts and the kind of material that erupts from it determine the shape of a volcano's cone.

What causes a volcano to erupt? Under every volcano lies a deep chamber of red-hot molten rock called **magma.** Pressure from hot gases forces the magma up to the surface, where it is then called lava. These hot gases are mostly water vapor. But magma also contains other gases, such as carbon dioxide and sulfur dioxide. Sulfur dioxide makes erupting volcanoes smell like rotten eggs. Gases, lava, and particles may all come out of volcanoes.

The drawing on the following page shows the three types of volcanoes.

- Volcanoes that don't erupt explosively form **shield volcanoes.** Lava flows out from the volcano's large opening and from surrounding cracks in the ground. When the lava hardens, it forms a very broad cone with gentle slopes.

- Volcanoes that erupt explosively form **cinder cones.** Notice that this kind of cone has steep sides. When this kind of volcano erupts, it throws out cinders (pieces of hardened lava) and large amounts of ash (very tiny particles). The cinders and ash form the cone.

- A **composite cone** is much larger than a cinder cone. It forms when a volcano erupts explosively at some times and quietly at other times.

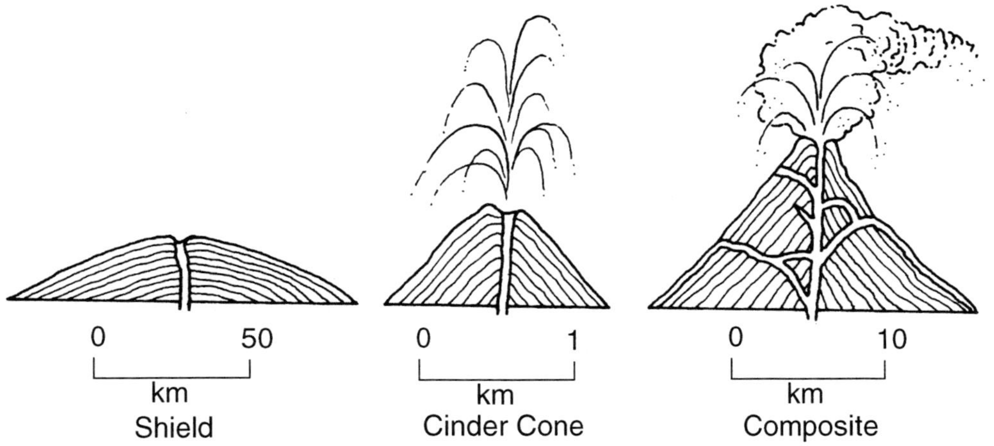

There are 600 to 800 active volcanoes and more than 1500 that have erupted in the last 10,000 years. Where are most of the earth's big volcanoes? Geologists have known for 100 years that many are located around the Pacific Ocean. They call this area the **Ring of Fire.** More than 75 percent of the earth's active volcanoes are on that border. Several others occur along the Mid-Atlantic Ridge, especially on and near Iceland. Others are present in east Africa along the Great Rift valleys.

The following map shows the locations of most major volcanoes. It also shows the earth's tectonic plates. Use the map to help you answer the questions on the following page.

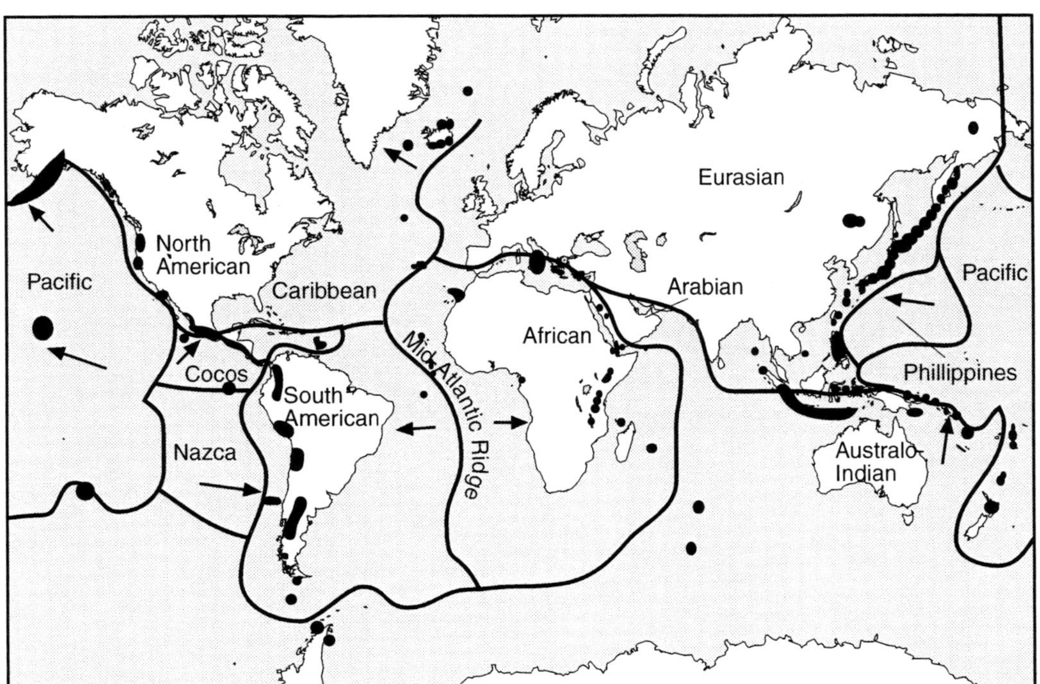

1. Where are most of the earth's volcanoes located?

____

____

2. How do you think volcanoes are related to plate tectonics?

____

____

____

____

____

3. Where are some volcanoes that are not located on plate boundaries?

____

____

____

____

____

4. From the map, what evidence do you see of a spreading plate boundary forming in east Africa?

____

____

____

____

Lesson 8

What Is Down There?

In 1864, French science fiction writer Jules Verne wrote the book *Voyage au Centre de la Terra* (*Journey to the Center of the Earth*). The book features German geology professor Otto Lidenbrock, his nephew Axel, and his assistant Hans.

The professor had found a centuries-old message written in an Icelandic runic alphabet. It read:

> "In Sneffels Yoculis craterem kem delibat
> umbra Scartaris Julii inta calendas descende
> audas viator, et terrestre centrum attinges.
> Kod feci. —*Arne Saknussemm*"

Axel finally decoded the message as follows:

> "Descend into the crater of Snaeffels Jökull
> that lies under Scartari's shadow just before
> the calends of July, audacious traveler,
> and you will reach the center of the earth.
> I did it. —*Arne Saknussemm*"

The three adventurers entered the crater of dormant volcano Snaeffels in Iceland and descended to the center of the earth. Among other exciting things, they saw an underground ocean complete with "living fossils," such as plesiosaurs and ichthyosaurs. They saw mushrooms 40 feet tall and a 12-foot-tall hairy giant herding Ice Age elephants (mastodons). After many hardships, they rode a raft of partly petrified wood up a volcanic vent. And where did their journey end? They are poured out of Stromboli, a continuously active Italian volcano.

The story is fantasy. But, it *is* possible to "see" what the center of the earth is like. To do this, we use information provided by two types of earthquake waves: **P waves** and **S waves.** As these shock, or seismic, waves penetrate the earth, they travel at different speeds. Recall from Lesson 6 that the P wave travels faster. The P wave is the primary wave. The slower S wave is the secondary wave.

The P waves push and pull the rock particles in the direction of wave movement. The S waves shake the rock particles at right angles to the direction of movement. (Imagine a rope tied to a tree with you shaking the other end up and down.) The back-and-forth motion of particles when a P wave passes means that each particle gives the particle next to it a push. This can happen as easily in a liquid as in a solid.

With the S wave, the motion requires that each particle drag along the next particle. This is not possible in a liquid or in a gas, where the particles are not attached to their neighbors as they are in a solid. So, S waves cannot travel through a liquid.

Both waves move fastest through denser (heavier) rock. Because the time the waves take to go through certain parts of the earth varies, scientists know the type of rock must vary. Also, both waves can be reflected or refracted (bent) at a boundary between two types of rock.

Scientists also know that the earth has a density of 5.52. (It is 5.52 times as heavy as an equal volume of water.) There are two main types of meteorites: metallic and stony. Meteorites are from our solar system. So it seems logical that the earth's interior could contain these same kinds of materials. All of these data together have been used to construct the model of the earth shown below. The **crust** of the earth is 3 to 40 miles thick, the **mantle** is 1788 miles thick, and the **core** is 2161 miles thick from the mantle boundary to the earth's center.

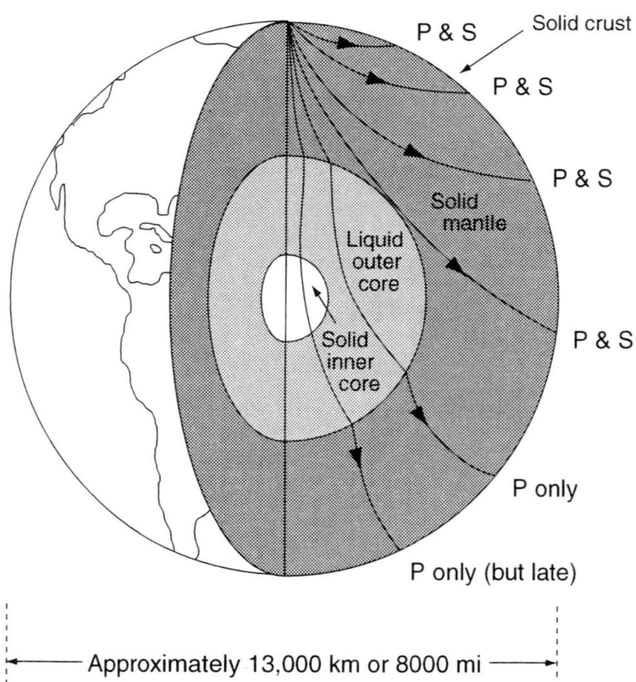

1. What evidence indicates that part of the earth's core is liquid?

____

____

____

____

____

2. How do the different layers on the earth model compare in thickness?

____

____

____

____

____

____

3. If you had to compare the earth's interior to a fruit, would it be more like an orange, a peach, or an apple? Explain your answer.

____

____

____

____

____

____

____

Lesson 9

The Making of Fossils

The picture below shows a baby woolly mammoth. The last of the woolly mammoths became extinct about 11,000 years ago. How did we get these pictures? The baby woolly mammoth was found preserved in frozen ground in Siberia. The animal still has hair on its legs. The woolly mammoth is an example of a **fossil**, the remains or traces of an organism that was once alive.

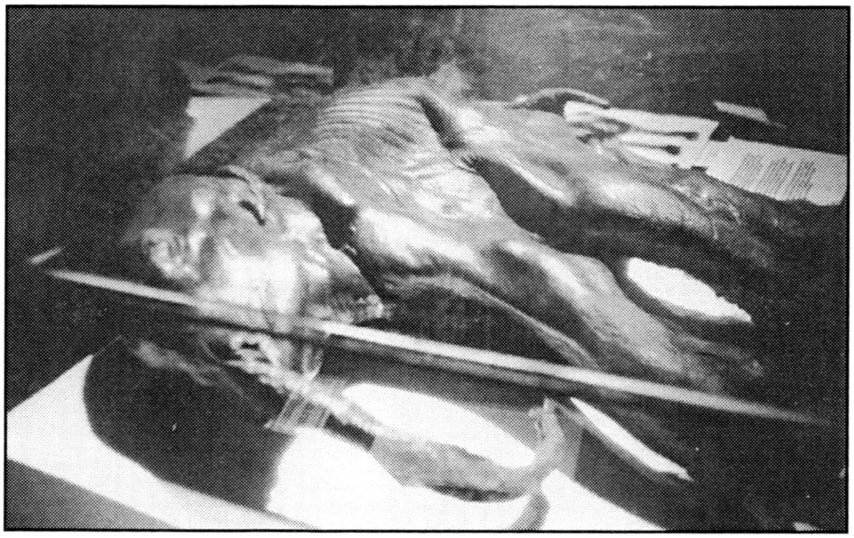

Woolly mammoth fossil

Hundreds of years ago, fossils were explained in weird ways. Some people thought that they grew in the rock. Others said that they were works of the devil. Today we know that fossils are traces, or remains, of prehistoric life. Not all fossils are remains of extinct animals. Some species living today, such as horseshoe crabs, are also found as fossils. The fossils tell us that those species also lived in prehistoric times.

There are two general requirements for fossils to form. The first is usually that the organisms have hard parts, such as teeth, bones, or shells. The second is quick burial before the organism is destroyed by weathering, decay, or being eaten by animals. In some cases, the original hard parts are preserved just as they were. Usually, however, the original hard parts are replaced by another material. For example, petrified wood is made of the mineral silica (SiO_2) that has replaced some of the tissues of the wood.

Most petrified wood occurs with volcanic ash. As the ash weathers, silica is freed and water moves it through the sediment. Many shells are replaced by calcite. A few are replaced by pyrite.

Rarely, soft parts can be preserved. Extinct Ice Age woolly mammoths and steppe bison that lived 10,000 years ago have been found frozen in Alaska and Siberia. Some, like the baby mammoth, had their hair preserved. And some even had green grass in their mouths.

Leaves can be preserved as delicate imprints. Insects become trapped in the sap of trees and are preserved when the sap turns to **amber.** Tracks, trails, and burrows made in mud and sand are found in rocks. Even fecal matter is sometime preserved.

Most organisms, like worms and insects, have soft bodies. Their chances of preservation, even as an imprint in mud, are very slight. Probably only a fraction of one percent of species that have ever lived are preserved as fossils. And many of these are destroyed by erosion and other changes in the rocks.

Millions of fossils exist in museum collections and in universities. Tens of thousands of species have been described. No one knows how many species of plants and animals are living on the earth today. Maybe tens of millions. There may be 30 million beetle species in the tropics. How many millions of species have lived and have then become extinct?

Fossils in layers of sedimentary rock led to the development of the geologic time scale. Geologists of 300 years ago realized that the fossil record showed that many life forms had become extinct. And they discovered that when an animal or plant became extinct, it never reappeared in younger layers of rock. They also realized that new life forms appeared in the rocks. All of this was early evidence for **evolution,** the changes in life through time.

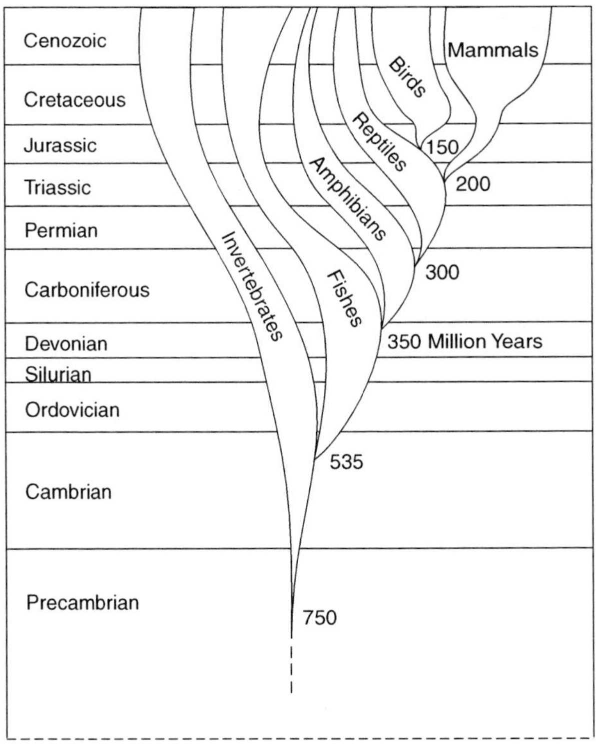

Major groups of vertebrate animals which appear in an order that supports the theory of evolution

1. Describe three different kinds of fossils and how they are formed.

2. Which would more likely form a fossil: a fish, a worm, or an insect? Explain your answer.

3. What can scientists learn from fossils?

4. During which periods did reptiles and amphibians first appear? Which type of animals evolved first?

Lesson 10

Death of the Dinosaurs: Why?

The first dinosaur fossils were found in England in the 1830's. They consisted of a few teeth and bones. These bones naturally caused much excitement. No one had ever seen such big bones before. That dinosaur was named Iguanodon. About the same time, other fossil teeth and bones were found elsewhere in England. These were named Megalosaurus. Soon dinosaur fossils were discovered in other parts of Europe and North America.

In 1841, Sir Richard Owen first used the term dinosaur to refer to the animals from which these large fossil bones had come. The word dinosaur means "terrible lizard." But as scientists later discovered, not all dinosaurs were large and terrible.

Then numerous big bird tracks were discovered in Massachusetts. These were soon recognized as dinosaur footprints. In the late 1800's and early 1900's, a great period of dinosaur hunting went on in the rocks of Colorado and other western states. In recent years, dinosaur track pathways have been interpreted as dinosaur migration routes. And large clusters of dinosaur nests with eggshells and even whole eggs have been found.

We now know that dinosaurs were the dominant life form on the earth for about 150 million years, through the Triassic, Jurassic, and Cretaceous Periods. Why then did dinosaurs disappear about 65 million years ago? There have been numerous hypotheses proposed to explain the **extinction:**

- The climate became so cool that continental glaciation occurred.
- Volcanism may have increased the CO_2 content of the atmosphere, causing a greenhouse effect which warmed the earth too much.
- Vegetation changed, affecting the entire food chain including carnivores.
- New diseases appeared.
- The atmosphere changed, from oxygen rich (35 percent) to the present 21 percent and the animals suffocated.
- Small mammals ate dinosaur eggs.
- New flowering plants caused severe pollen allergies.
- A supernova exploded near the earth.
- Dinosaurs ate too many green plants, got diarrhea, slid into the ocean, and drowned.

Few of these hypotheses have enough evidence to support them. A more scientific hypothesis was suggested in the late 1970's by Walter Alvarez, a geologist, and his father Luis Alvarez, a physicist. With two chemists, they discovered abnormally high amounts of the rare element **iridium** in a thin clay layer at the boundary between Cretaceous and Tertiary rocks in Italy. They suggested the iridium came from an asteroid impact at the time of dinosaur extinction. Most scientists doubted the hypothesis, but it was the spark for much cooperation between different branches of science.

Since the early 1980's, iridium has been found at the same Cretaceous/Tertiary (K/T) boundary in more than 100 other places in the world. And a buried crater over 100 miles in diameter has been found in Mexico. It has been interpreted by impact enthusiasts as the **impact crater** of an asteroid or a comet. During impact, huge quantities of dust were thrown high into the atmosphere. This blocked out sunlight and the earth was darkened for months, or even years. Plants couldn't grow and animals starved. Some charcoal found at the K/T boundary suggests that worldwide fires were caused by the impact.

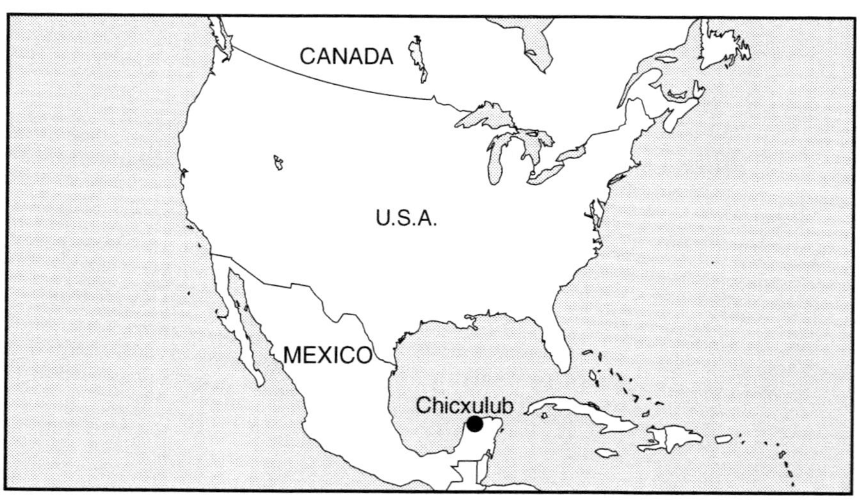

Impact crater site at Chicxulub, Yucatan Peninsula, Mexico

Opponents of the impact hypothesis say that the iridium could have come from the earth's mantle during volcanism. Similarly, much other "impact evidence" has been interpreted as the result of volcanism. In fact, much volcanic rock was formed in India and Brazil at that very time.

Some paleontologists say that the extinction of dinosaurs and other forms of life occurred gradually over time. They say the fossil record is so incomplete that the extinction just *appears* to have been sudden.

Use reference sources, such as library books and magazines or the Internet, to find information to help answer the following questions.

1. Suppose plant-eating dinosaurs became extinct when changes in climate affected plant growth. How would their extinction affect the meat-eating dinosaurs?

____

____

2 Were dinosaurs the only life forms that became extinct at the end of the Cretaceous Period? Explain.

____

____

3. Besides the extinctions that occurred about 65 million years ago, there have been other major extinctions. When did these occur? What kinds of life became extinct?

____

____

____

____

____

____

4. What other kinds of reptiles that lived during the Age of Dinosaurs survived the extinction at the end of the Cretaceous Period?

____

____

Lesson 11
The Ice Is Moving

Imagine traveling back in time 15,000 years. If you could view North America from space, it would look very different. A huge sheet of ice covers all of Canada and the northern part of the United States. There are no Great Lakes. The central United States is not covered by prairies but by dense spruce forests. The states of Utah, Nevada, and California have many large lakes. Another large lake covers most of southern California. You would be seeing what North America looked like during the last Great Ice Age when much of the land was covered by glaciers.

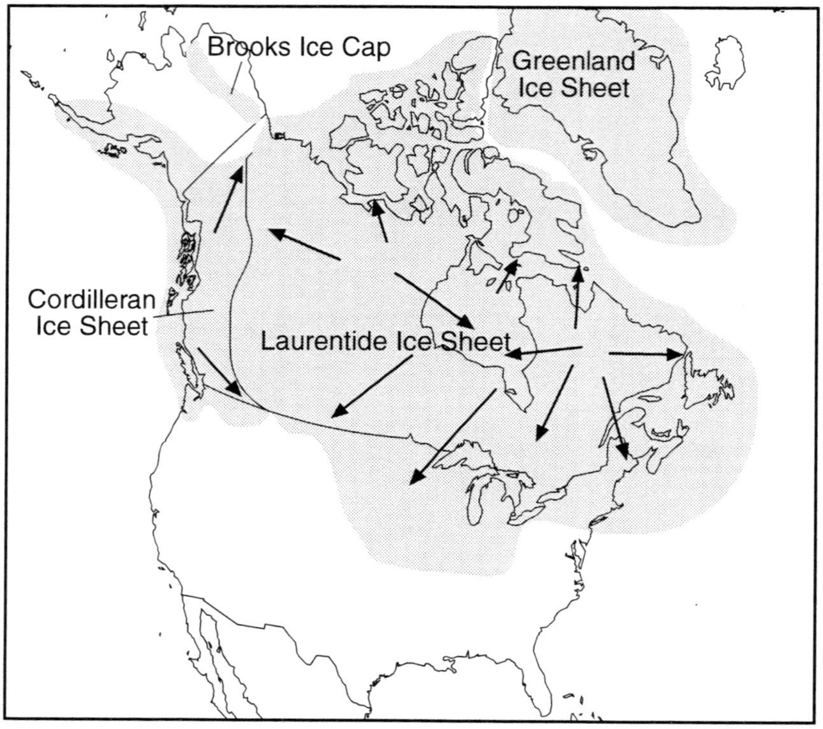

Map showing the extent of the continental ice sheet on North America during the Great Ice Age of the last two million years

Today about 10 percent of the earth's land surface is covered by glaciers. During the height of the Great Ice Age, three times as much was covered. How do glaciers form? If more snow falls each year than melts, the snow builds up. By the time the snow is 200 feet thick, its weight will have turned the lower snow into dense ice. Then the weight will cause the lower material to behave like plastic and flow.

Ice flows like liquid water, but much more slowly. It moves or flows downhill, under the influence of gravity. Based on size and location, glaciers are of two main types. One type is **valley glaciers.** They develop in high mountain valleys and flow like rivers of ice slowly down the sides of the mountains. Thousands of them exist today.

It is easy to understand why a glacier in a mountain valley moves. It is simply moving downhill like water, under the influence of gravity. A glacier picks up stones and sand that act like sandpaper as the glacier moves over solid rock. The result is grooved and polished rock. A mountain glacier will transform a V-shaped valley eroded by a river into a U-shaped valley. The valley shape is one clue that tells a geologist that a mountainous area was once covered by glaciers. Other clues are bowl-like features called **cirques** at the upper end of a glacial valley. A cirque is the starting point of a valley glacier. A mountain peak may have cirques on several sides, making it a horn. Sharp ridges (aretes) between adjacent glacial valleys are still further evidence of glaciation. A pile of poorly sorted rubble at the end of a glacial valley is a terminal **moraine.** It is made up of material eroded from the valley walls and carried by the glacier. When the glacier flows down to a lower and warmer elevation, it melts and deposits its load as a moraine. Smaller glacier valleys joining bigger valleys are called hanging valleys.

The other type of glacier is the **continental glacier** or **ice sheet.** Today ice sheets exist only in two places—Antarctica and Greenland. A continental ice sheet is very large, and also moves. But why does it move, since it is not necessarily moving downhill? An ice sheet can become as much as two or three miles thick. The weight of the ice itself causes the glacier to flow out from its thickest part. Ice at the bottom of a glacier melts because of pressure. Water from the melted ice lubricates the glacier and helps it move. In this way, the continental ice sheet spreads outward. When it reaches a warmer latitude, it no longer erodes the bedrock. Instead, it melts. If a sudden climactic warming occurs, the glacier can stop moving and melt. Or, the ice may keep moving forward but not advance beyond a certain point because its lower end keeps melting. This can result in a ridge (terminal moraine) of unsorted material, called **till.** Meltwater running off in front of the glacier will sort the material into beds of sand gravel, and boulders.

At many places in Canada, the midwestern United States and in New England, there is scratched, and grooved, and polished rock. There are hills made of unsorted material. There are flat areas with scattered lakes.

There are scattered boulders of rocks types from distant northern areas. This evidence says that continental ice sheets advanced and melted back four times in the last two million years.

Use the information in this lesson and reference sources such as library books and magazines or the Internet to answer the following questions.

1. Label the features on the diagram below that tell a geologist the area has been glaciated. Use the terms arete, cirque, glacier, hanging valley, horn, and moraine.

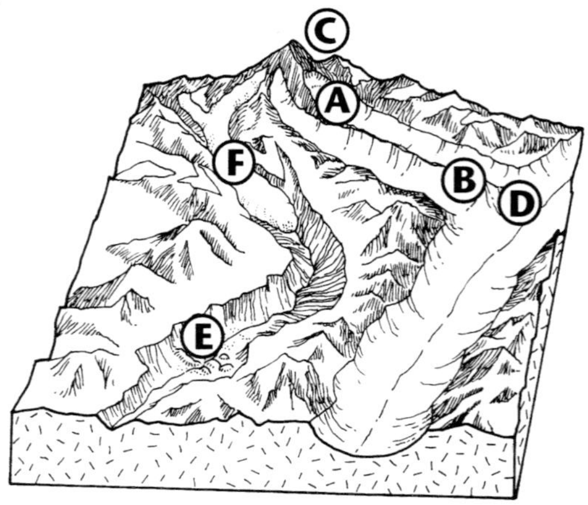

2. What are some ways that glaciers have changed the landscape of the United States?

____

____

____

3. Suppose you found two rocks, one smooth and rounded and the other with one side flat and covered with scratches. Which pebble is more likely to have been moved by a glacier? Explain your answer.

____

____

____

Lesson 12

Soil: Earth's Valuable Resource

If you reach down and pick up a handful of soil, you will be holding one of the earth's most valuable resources. Soil provides a place for plants to grow and home for millions of other organisms. Where does soil come from? Soil is formed from rocks.

Some igneous rocks and all metamorphic rocks are formed deep in the earth's crust. Volcanic rocks crystallize from hot magmas. When these rocks are on the earth's surface, they come in contact with air and water. Then the minerals in the rocks change to other minerals. The main new mineral when soil forms is clay.

The changes in the rocks are due to two weathering processes—physical and chemical. **Physical weathering** (also called mechanical weathering) includes the processes whereby rocks and minerals are broken into smaller pieces. This provides more surface area for chemical weathering. **Chemical weathering** is the chemical breakdown of rock, with new products (mainly clay) formed in the process.

Rain water, groundwater, and surface water are slightly acidic. Acid water causes changes in rock faster than pure water does. Any minerals containing aluminum, silicon, and oxygen, especially feldspars, can be changed to clay.

Climate is the main factor that determines the type of clay mineral formed. Heat and moisture speed up chemical reactions. In a polar climate, chemical weathering does not occur. In a subtropical or tropical climate, where it is hot and wet, chemical weathering occurs too quickly. The elements necessary for a good soil (potassium, sodium, calcium, magnesium and many others) are dissolved and carried to the oceans. The soil in such a climate is made up of a simple clay that contains hydrogen, aluminum, silicon, and oxygen. This soil is red because of iron oxide (iron combined with oxygen).

But a **temperate climate** is just right. It is neither too cold, too hot, nor too wet. There the elements essential for plant growth stay and become part of the clay. These clays contain the elements potassium, sodium, calcium, magnesium, aluminum, silicon,and oxygen.

A soil contains clay minerals, iron oxides, some grains of stable quartz and incompletely weathered minerals, and some decayed organic matter from plants and animals called **humus.** Water and air are also present. The drawing shows soil that has developed on weathered rock.

Why is a good clay so important in soil? Because plants can easily remove essential elements (nutrients) from clay. These elements are loosely held in the clay minerals. These nutrients are so strongly bonded in the crystal structures of minerals (such as feldspar) that plants cannot easily remove them.

So the best soil develops in temperate climates. High-protein crops such as wheat (10 to 14 percent protein) grow in such soil. Food crops can be grown in tropical or subtropical soil, but they contain less protein. For example, bananas, which grow only in tropical areas, have only 3 percent protein. And proteins are important for building strong bodies.

Write the answers to the following questions in your notebook.

1. Where is the best soil for growing food? Why?
2. Why are good clay minerals in soil so important?
3. How are mechanical and chemical weathering different?
4. Explain how parts of the soil in your garden might have once been part of a mountain or a tree.
5. How do you think farming has changed the nature of soil around the world?

Lesson 13

Where Does the Water Go?

It's raining for the third day in a row. Where does all that water come from? And where does it go? The water in the rain falling on your yard constantly moves in a cycle. The water cycle, or **hydrologic cycle,** shown below is constantly in action. It collects, distributes, and even purifies water all over the earth. It is one of the most important cycles for sustaining life.

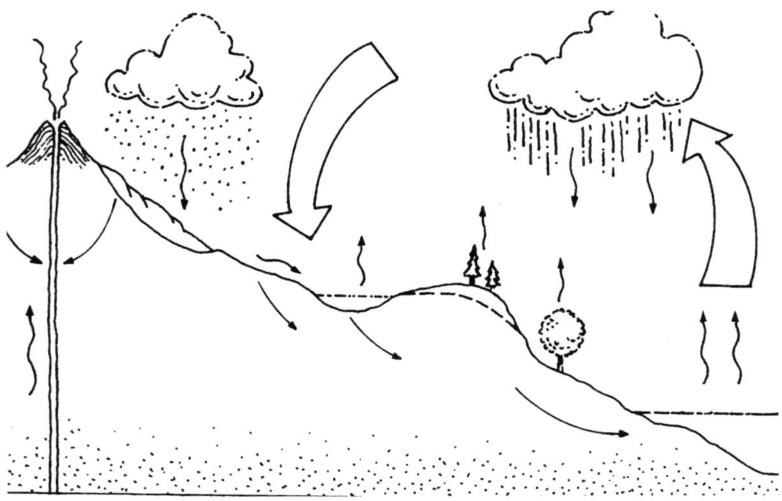

Hydrologic cycle showing routes that water takes as it is recycled over and over again

After rain or snow, some water soaks into the soil and ground. This is called **infiltration.** Some runs off and flows downhill under the influence of gravity. What controls whether water runs off or soaks in? A slow rain over land that is covered with plants allows much of the water to infiltrate. But what if too much rain falls too fast, and the soil and underlying rock are already **saturated,** or full of water? Then the water must run off in a downhill direction.

The ultimate destination of water is the ocean. However, an individual drop of water may not make it there on this trip. It may evaporate from the ground as it flows downhill or from a stream or river's surface. Then, once again it will fall as rain or snow.

In a given area, rain falls over the entire land surface. This thin layer of water, called **sheetwash,** is responsible for erosion of the land. As the sheetwash layer flows toward lower areas, it erodes land. It forms trickles of water and the trickles join to become creeks. The creeks join to become rivers, and the rivers join to become larger rivers.

Maybe you thought rivers cause erosion? Indeed, they can erode land along their channels. However, the main thing rivers do is act like freight trains that carry water out of an area. For example, the Mississippi River has a tremendous flow of water, especially at **flood stage.** This water also transports one to two million tons of fine-grained **sediment** into the Gulf of Mexico each day. This compares to loaded ore trains, one after another, driving into the Gulf 24 hours each day.

All of this connected water—the sheetwash, the trickles, the creeks, and the rivers—drains a specific area of land called a **drainage basin.** Crests of high land, called divides, separate different drainage basins.

Most large rivers that are not in the mountains where the slope of the land is steep, have a **floodplain.** A floodplain is a flat, low-lying part of a river valley on both sides of the river channel, shown in the diagram below. As the name implies, the floodplain is covered by water when a river floods or overtops its channel. Rivers commonly meander or wind across their floodplains. Some floodplains may go decades without being totally covered by water. Specialists speak of 10-year floods and 100-year floods. During these floods, water levels rise way above normal flooding.

River valley showing the river channel and the floodplain

Sometimes people build homes or even towns on floodplains. They may never have seen the land flooded, so it may seem safe. But if too much rain falls too fast, or if a winter's snowfall melts too rapidly on frozen ground, floods are the result. Some can be flash floods, large floods when streams and rivers rise very fast. These floods can be dangerous because people don't have time to escape the rising water.

1. Study the diagram of the hydrologic cycle. Then, describe one possible way a single drop of water might make a complete trip through the water cycle.

____

____

____

2. Name some natural factors that affect runoff. Name some human activities that would affect runoff.

____

____

____

3. Why do you think a floodplain is usually a good farming area?

____

____

____

4. Why is a very heavy rain more dangerous than a long, light rain?

____

____

____

Lesson 14

Water, Water, Everywhere

"Water, water, everywhere, nor any drop to drink," is a line from Samuel Taylor Coleridge's 1798 poem *The Rime of the Ancient Mariner.* Imagine being on a sailing ship in the middle of the Pacific Ocean near the equator with no wind. The ship is stranded; the crew is hot and thirsty. There is water all around, but it is salty and undrinkable.

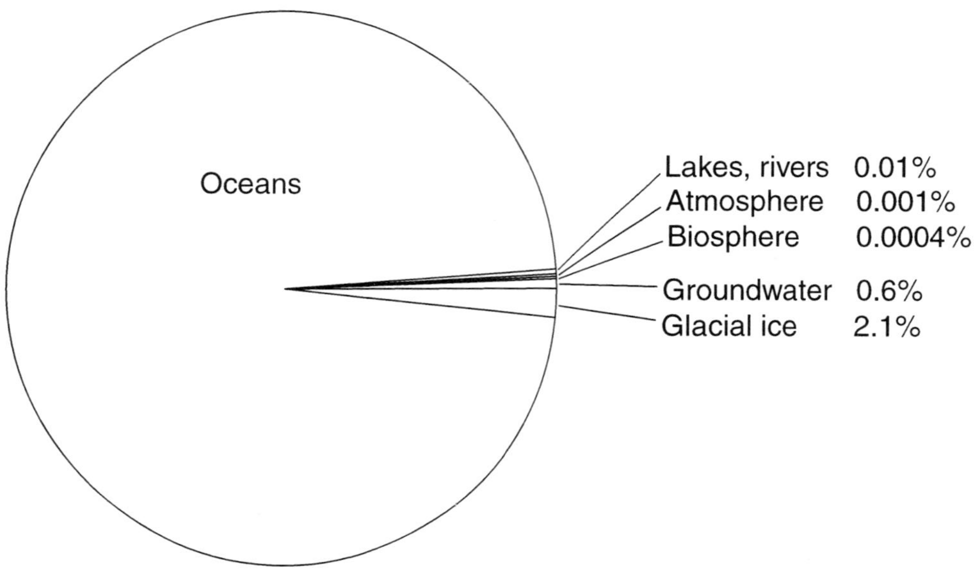

The earth's water

The earth is the water planet of our solar system. About 71 percent of the earth's surface is covered by water. There is no shortage of water on the earth. But how much of that water is useable? The earth has an estimated 359 quintillion gallons of water (1×10^{18}) gallons. Of this, 97.3 percent is in the oceans. Ocean water is salty and cannot be used for drinking or farming. The remaining water on the earth is fresh water. About 2.1 percent of the earth's water is frozen in glaciers, mostly in Antarctica and Greenland. Thus it is not very reachable. Only 0.6 percent of the earth's water is **groundwater,** water that is underground in the soil and rocks. And only 0.01 percent is in lakes and rivers. So, only 0.61 percent of the earth's water is easily useable. We indeed seem to have a shortage of fresh water.

So, much less than 1 percent of the earth's water is readily available for human use. What do we use it for? The graph to the right shows how water is used in the United States. But this graph doesn't give the whole story. For example, in the western half of the United States, 85 percent is used for irrigation. And much of this just evaporates before it can be used by plants. Growing one pound of cotton requires 2000 gallons of water. Raising one pound of grain-fed beef requires 800 gallons. Growing a pound of rice requires 560 gallons. To make one loaf of bread, 150 gallons are used along the way. The manufacture of one automobile requires 100,000 gallons of water.

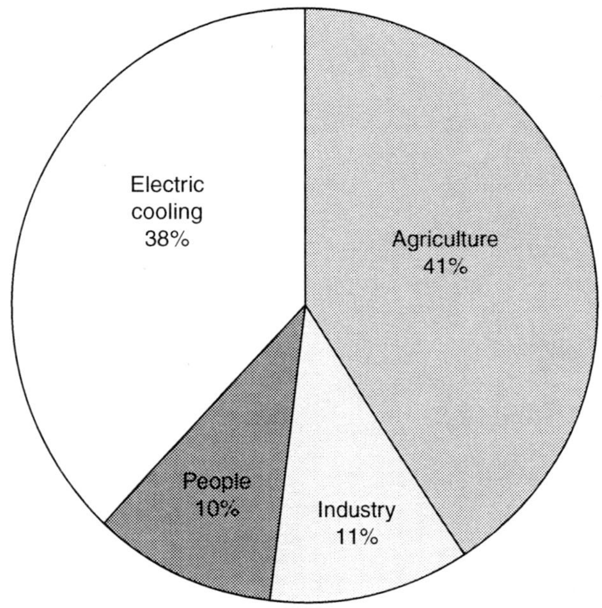

Water use in the United States

Nearly a quarter of the water used in the United States is groundwater and comes from wells. Many water-bearing rock units underground, called **aquifers**, are being depleted. As you might guess, fresh water is especially scarce in semi-arid and arid areas. In such dry areas, groundwater recharge is naturally slow because groundwater moves very slowly. We are pumping it out faster than new water can soak into the ground. Groundwater is also being polluted by industrial, agricultural, and human activities.

Droughts, long periods of dry weather, further increase water shortages. Droughts have been especially bad in Africa over the last few decades. No water means no crops, and no food for people or animals.

People get most of their water from rivers and lakes. They build dams to make reservoirs of fresh water, capturing water as snow melts, or from run-off during the rainy season. A dam creates a large new lake. The lake has a large surface from which water can evaporate. And the reservoir can silt up because the river feeding it carries in fine sediment. As humans try to control nature, new problems always arise. And people pollute rivers and lakes, as well as groundwater. The Clean Water Act in the U.S. has been in effect since 1972. Sewage treatment plants are one result of the act. But there are still pollution problems that must be dealt with.

Use reference sources, such as library books and magazines or the Internet, to find information to help you answer the following questions.

1. List as many ways you can think of about how your family uses water every day. What could you do to reduce the amount of water your family uses?

____

____

____

____

____

____

2. List one advantage and one disadvantage of using dams to help manage our water supply.

____

____

____

3. One future answer to the fresh water shortage may be desalinization. What is desalinization? What areas of the world would benefit most? What are its problems?

____

____

____

____

____

Lesson 15

Way Down South

A few hundred years ago, world maps included a large southern continent called *Terra Australis Incognita,* or the "unknown south land." But the continent's presence was suspected long before that time by the early Greeks. They thought a very large southern continent was essential to "balance" the big northern continents. The Greeks named it Anti-Arktikos, opposite of the Arctic.

Antarctica wasn't actually sighted until 1819 when William Smith's ship was blown south to what are now the South Shetland Islands. British, Russians, and Americans sailed to the region before 1830, and some sailed all the way around Antarctica. Others followed.

The early explorers noted the abundance of life in Antarctica's waters—whales, seals, penguins, plus countless seabirds. Soon the sealers and the whalers arrived. Thus began one of the most inhumane chapters of history, the slaughter of seals and whales.

The seals were not afraid of humans. More than 1.2 million Antarctic fur seals were killed between 1800 and 1822. By 1912 when sealing ended, three million fur seals had been killed. Another one million elephant seals were killed for their oil. More than 1.3 million whales have been killed since 1904. Only 1 percent of the big blue whales, 3 percent of humpback whales, and 30 percent of fin whales survived.

Fortunately, Antarctica became a whale sanctuary in 1994. Today, some seals, fish, and **krill** are harvested by Russia and Japan, but with strict quotas. Krill are small, shrimplike animals that are the main food source for fish, birds, and seals. Scientists are concerned that this loss of krill will affect the food chain.

In spite of the abundance of life in Antarctic waters, the interior of Antarctica is almost lifeless. The life is along the coast and islands, and then only during Antarctica's summer. A whopping 98 percent of Antarctica is covered by glacial ice one to three miles thick. The glaciers are continually moving off the continent toward the ocean. Therefore, hidden crevasses are a constant danger to geologists. The geologists are slowly piecing together the history of the continent by studying the exposed rocks that comprise the remaining two percent of the surface. Underneath its ice,

Antarctica has mountains, valleys, and lowlands, just like the landforms of the other continents.

East Antarctica is the main part of the continent. West Antarctica consists of islands covered by a marine ice sheet, most of which is "grounded," or sitting on the ocean floor. The three large **ice shelves** on the coasts are huge masses of floating glacial ice that shed massive icebergs. One iceberg was as big as the state of Rhode Island.

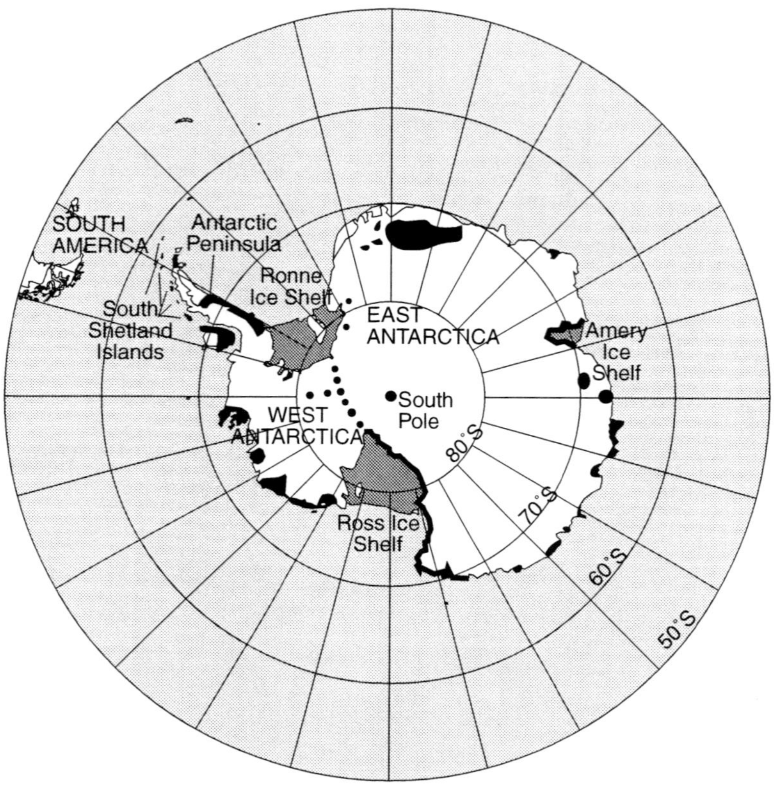

Continent of Antarctica showing exposed rock as black areas, glacial ice as white areas, and ice shelves as dark gray areas

Antarctica makes up about 10 percent of the earth's land surface and it is surrounded by the world's stormiest ocean. The westerly winds blow around Antarctica, unhindered by any other landmass. The 600-mile-wide Drake Passage between the tip of South America and the Antarctic Peninsula is especially rough, with huge waves.

Use reference sources, such as library books and magazines or the Internet, to find information to answer the following questions.

1. Who where the first two explorers to reach the South Pole? Tell about what happened on their journeys.

____

____

____

____

____

____

2. How did the Antarctic icecap form? How could it form if Antarctica is considered a desert?

____

____

____

____

____

3. What is the main difference between east and west Antarctica? What separates the two regions?

____

____

____

____

____

____

Lesson 16

Take a Really Deep Dive

Imagine that you have graduated from one of the country's leading oceanographic institutes and are now working as a marine geologist. You are on board the *Alvin,* a 25-foot-long mini-submarine that can dive to depths up to 15,000 feet. Lucky for you, the *Alvin* is built to withstand great pressures. The six-foot sphere is made of six-inch-thick titanium metal. There are four portholes, one on the front, one on each side, and one on the bottom. The plexiglass portholes are six inches thick.

Drawing of *Alvin*

You share the cramped space with a marine biologist and the pilot. There is no heat in the *Alvin.* The only heat is generated by your three bodies. There's no bathroom, either. An oxygen bottle supplies essential oxygen for breathing and the carbon dioxide you exhale is automatically removed.

Today's dive will be only to 8200 feet. You will be underwater for about eight hours. It will take an hour and a half to get down to that depth and the same amount of time to get back up. That will leave you about five hours to make observations, take photographs, and collect samples. You will be using *Alvin's* two mechanical arms to collect volcanic rocks. The marine biologist will collect animals.

After descending for an hour and a half, the *Alvin* is a mile and a half deep, on the ocean floor. At this depth, it is pitch black outside and the pressure is 250 times as great as at the ocean's surface. You think, "I sure hope that six inches of titanium is strong enough." The place you are exploring is on the East Pacific Rise, the **spread center** of the Pacific Ocean. Here new volcanic crust is being formed. Old flows of cold volcanic rock are all around you, forming mounds 10 to 12 feet high. They are made of **pillowed lava**, typical of underwater flows.

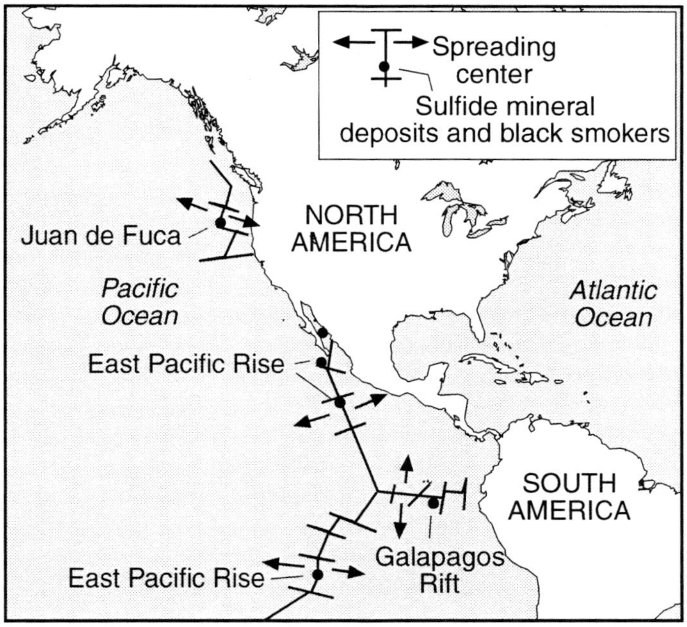

Then you see it, illuminated by *Alvin*'s lights. A **black smoker** is just ahead. Hot, dark water is being spewed out of a vent at the top of the structure. The black smoker chimney is about 20 feet tall and 3 feet in diameter. It contains many metals, such as iron, copper, and zinc. The water coming out of the vent and nearby holes has a temperature of about 400° Celsius. The rising hot water is ocean water that moved down through basaltic volcanic rocks, became heated, and then rose. This is **hydrothermal** activity, and sulfide ore deposits are being made right before your eyes!

What are those funny-looking white tubelike creatures with red tops? They seem to be waving at you through the portholes, as they move back and forth in the water above their attachments to the rocks. The marine biologist tells you they are giant tube worms. The four-foot-long animals get their energy from sulfur in the form of hydrogen sulfide gas. They can't get their energy from sunlight because no light penetrates this deep. Then you notice crabs crawling around and a fine, white, snowlike material being expelled from the vents. The white material is bacteria that are the base of a food chain that is very different from that which we know on the earth's surface.

You have seen something very few people have ever seen—a strange underwater world interrupted by bursts of volcanism.

Use reference sources, such as library books and magazines or the Internet, to find information to help you answer the following questions.

1. Why are geologists interested in studying black smokers?

2. How does the food chain at the smokers differ from that near the ocean surface?

3. Why isn't the East Pacific Rise in the middle of the Pacific Ocean just as the mid-Atlantic Ridge is in the middle of the Atlantic Ocean?

Lesson 17

Strong Winds

Imagine that you live along the coast of North Carolina. It is early August. You have been listening to the radio and you know a **hurricane** is 300 miles away and approaching from the southeast. What would you experience? As you look up, you would see high cirrus clouds in delicate bands and wisps. The **barometer** is registering a slow drop in atmospheric pressure.

By the time the hurricane is 150 to 200 miles away, the winds are blowing 30 miles per hour, and increasing steadily. The hurricane continues to get closer. When it's about 100 miles away, the winds are 50 miles per hour and the barometer shows a very low pressure. The sky is full of dark, turbulent clouds.

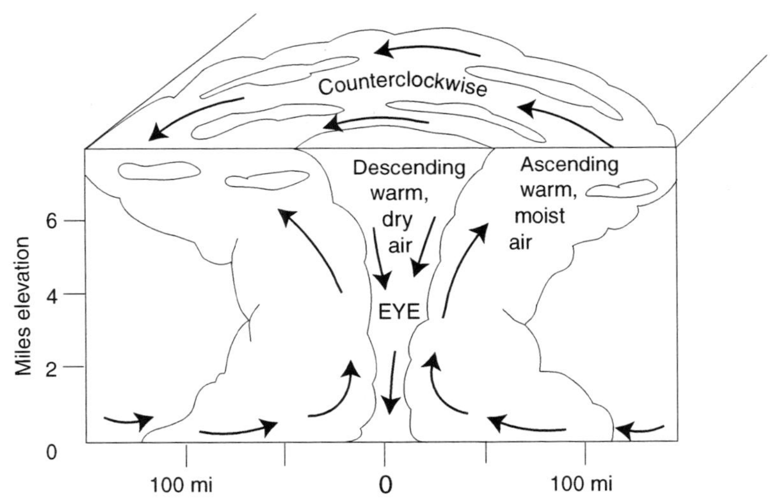

Cross-section of a hurricane

Now the hurricane reaches you. Winds top 100 miles per hour! Tree branches break, shingles come flying off roofs. Rain comes down in torrents from dark clouds. After several hours of this, the winds suddenly drop to 20 miles per hour. It feels warm. You even see the sun. But the air pressure is very low. The center of the storm, or **eye** of the hurricane, is passing over you. This relative calm lasts only about half an hour.

Then, as the other side of the hurricane begins to pass over, you experience the same high winds and heavy rain as before. Eventually, the winds and rain die down. Hurricane season has started again.

Now, imagine you live in central Kansas. It is late afternoon in the middle of June. A cold front is approaching from the north, meeting warmer and moister air that has moved northward from the Gulf of

Mexico. A line of thunderstorms has broken out along the boundary of the two air masses, right near your town. Suddenly the town sirens begin wailing! A **tornado** has been spotted. You scan the horizon. There it is!

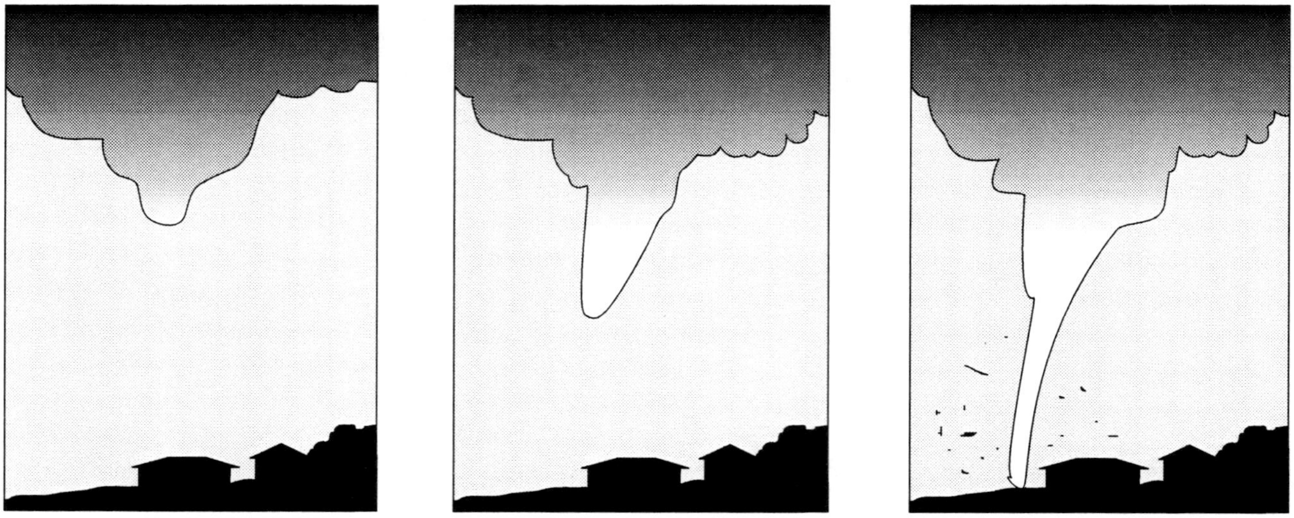

A funnel-shaped cloud is projecting downward from the dark, rolling clouds above. Slowly the funnel gets longer and narrower and then it touches the ground. You see debris flying all around it—dust, pieces of houses, and entire trees torn up by the roots. You head for your basement and wait for the tornado to pass. Fortunately, it misses your neighborhood. Later you hear that dozens of mobile homes and houses were damaged or destroyed. Many people were injured; several are missing and feared dead. Once again, it is tornado season in the Great Plains.

Use reference sources, such as library books and magazines or the Internet, to find information to help you answer the following questions. Write the answers in your notebook.

1. What is a typhoon? What is a tropical cyclone?
2. Why do waves get so large during hurricanes?
3. Why do hurricanes slow down and weaken over land?
4. How do the wind speeds in a hurricane and tornado compare?
5. Which kind of storm do you think is more dangerous—a hurricane or a tornado? Explain your reasoning.
6. What safety practices should you follow if you are in the path of a hurricane?
7. What safety practices should you follow during a tornado warning?

Lesson 18

Moving Beaches

Imagine walking barefoot along an ocean beach, picking up seashells. Or just lying on the white sand beach in the warm sun. If you have a favorite beach, you might think it will always be there for you to enjoy. But that isn't necessarily so because beaches are always changing.

There are two main types of beaches. One type is found along the shores of lakes or oceans, like those found along the Pacific coast of the United States. The other type of beach is found on **barrier islands.** Barrier islands are common along the Atlantic and Gulf coasts of the United States. In fact, North America has more than half of the world's barrier islands.

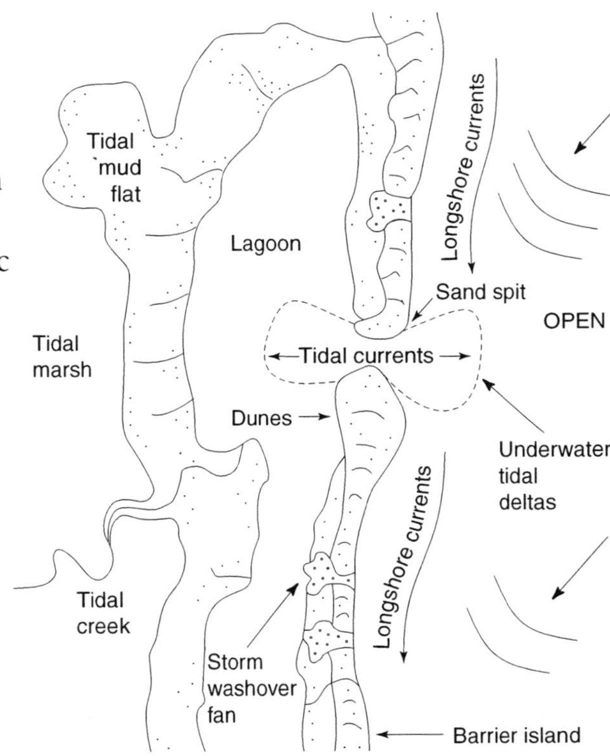

Barrier islands are long, narrow sand islands found just off the mainland. Barrier islands may be located a mile or more from the shore. The water that separates them from the mainland is called a lagoon. Sand can be washed inward over the islands during storms or through gaps in the islands.

Beaches have been called "rivers of sand" because the sand, both on shorelines and on barrier islands, is always moving. If waves strike a shoreline at right angles, the sand will move in two directions—in and out. But waves usually meet a beach at some other angle, and this creates a **longshore current.** This current carries sand down the shoreline.

Sand also moves differently during different seasons. One summer day, Richard's family discovered a small sandy beach along the Pacific coast of California. The next January, they decided that beach would be the perfect spot for a picnic. When they arrived, they thought they were at the wrong place. There was no sand at all. It was, instead, rocky with big rocks everywhere. What had happened? It was winter and strong winter waves had carried all the sand out to sea. By the next June, the area was once again a sandy beach. The gentler summer waves had slowly moved the sand back. Beaches grow during periods of calm seas.

Many people build homes along beaches even though they know the sand moves in and out and along the beaches. This movement happens even during normal weather. But during storms and hurricanes, sand *really moves*. Beaches with their loose sand grains are easily destroyed by **erosion.** Homes built on beaches topple as the sand is removed. In 1992, Hurricane Andrew removed almost 80 percent of the sand from Louisiana's barrier islands.

People want to preserve the beaches and prevent such damage. Can it be done? The U.S. Army Corps of Engineers has tried. Jetties, extending out into the ocean, have been built to protect harbors. Walls have been built perpendicular to beaches to try to keep the sand from moving away. Rock seawalls have been built to resist the waves. Artificial sand dunes have been built. Sand has been pumped from deeper water onto beaches to replenish the sand.

Human efforts have temporarily slowed beach erosion in some areas while making it worse in others. For example, a jetty built to protect an inlet bordering Ocean City, Maryland, blocked the flow of sand to Assateague Island. As a result, the island has thinned and migrated toward the mainland. The expense of maintaining barrier islands for development is very costly. A single year of bad storms can cause up to $4 billion in property damage. After a bad storm, taxpayers may be faced with paying for beach repairs.

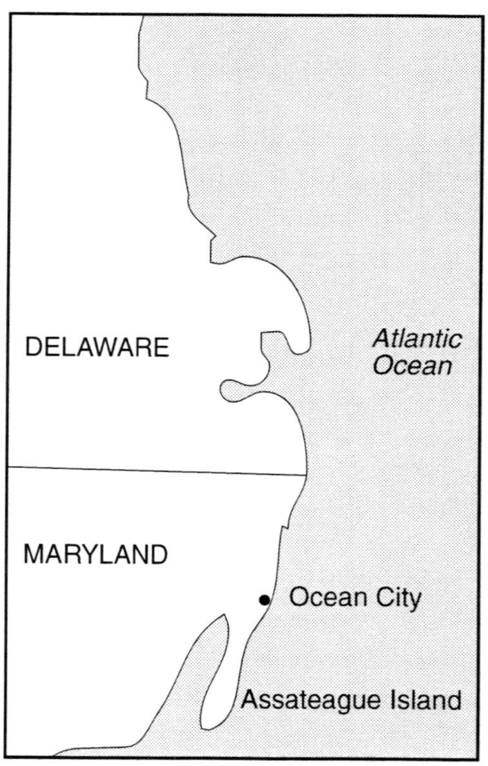

But, beaches are the land's defense against the sea. They absorb storm energy by changing their shapes. Then they rebuild during calmer weather. Human engineering interrupts the natural movement of these rivers of sand. It changes the natural behavior of beaches and it costs billions of dollars.

1. Why have beaches been called "rivers of sand?" Do you think this is good way to describe beaches? Explain.

2. Describe how and why beaches change.

Lesson 19

Lots of Energy

Can you imagine life without cars, trucks, buses, and airplanes? All these vehicles run on fuels made from oil. Estimates of how long the world's oil supply will last vary from 20 to 100 years. It is possible that the world will run out of oil in your lifetime. How could that affect you?

There are 550 million road vehicles in the world, about a third of them in the United States. More than half of the oil consumed in the United States is as gasoline to drive vehicles on the highways. In 1996, Americans drove 336 million miles each day. People use cars, trucks, and buses to travel to work, school, and many other places. But trucks also carry raw materials and finished products to all parts of the country. And airplanes and ships also need fuels made from oil. Without oil, how would people and products get from one place to another?

Transportation is only one way we use oil. Heating oil is used to heat homes and other buildings. We also use oil for agriculture. Tractors run on fuels produced from oil, and fertilizers and insecticides are manufactured from oil. Without oil, the United States could not be the world's biggest food producer. Oil products are also the basic materials used for the manufacture of synthetic fibers for clothing, plastics, paints, soaps, and synthetic rubber.

Oil, along with natural gas and coal, make up what are known as **fossil fuels.** Fossil fuels are fuels that formed from the remains of prehistoric life. The formation of fossil fuels we use today began millions or hundreds of millions of years ago. Because it takes so much time for these fuels to form, they are considered **nonrenewable resources.** When such resources are used up, they are gone for good.

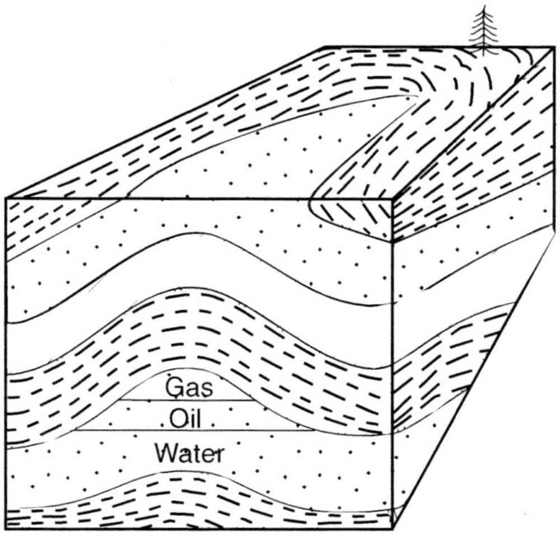

The energy in these fuels came from energy stored in the bodies of living things. Originally this energy came from sunlight in the form of solar energy. Layers of plants, when buried under layers of other sediment, can be converted into coal. Microscopic animals and plants, under the right conditions of pressure and temperature, can develop into droplets of crude oil or natural gas. This happens in black muds, the "source rock" for oil. Then, water moving through the sediment collects the oil and it can be trapped in a reservoir rock of sandstone or limestone. Once geologists locate this "pool" of oil or gas, it can be pumped out of the rock.

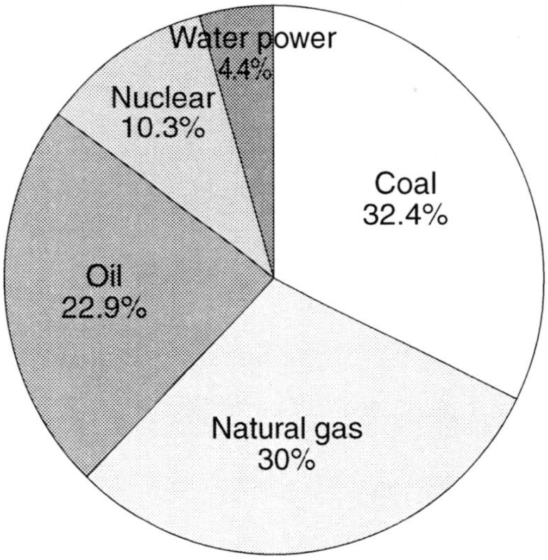

Currently oil, coal, and natural gas supply about 90 percent of the world's energy needs. The pie graph shows the energy sources used in the United States. While world oil supplies are not expected to last more than 100 years, it is estimated that natural gas supplies will last from 55 to 160 years and coal supplies may last for another 300 years. Coal and natural gas are used to generate most of the world's electricity. How will we produce electricity when these energy resources are gone?

Use reference sources, such as library books and magazines or the Internet, to find information to help you answer the following questions.

1. When and where was the first oil well drilled?

____

____

2. The United States uses about 6.5 billion barrels of oil per year. The total known U.S. oil reserves are about 25 billion barrels. How long will this supply last us? How long will it last if we found new supplies that quadrupled our reserves?

____

____

3. The world's known oil reserves are 1000 billion barrels. The world's population uses 50 to 60 billion barrels a year. How long will the oil last, assuming no increase in use and no new discoveries of oil reserves?

____

____

4. Look at the pie graph on the previous page. How much of our energy comes from fossil fuels?

____

____

5. What are some ways that people can help make the earth's supply of nonrenewable energy resources last longer?

____

____

____

____

____

____

____

Lesson 20

Earth Is a Spaceship

The earth is a spaceship moving through space on a long, long voyage. What does a spaceship need for a successful journey? Think before you read further. Then, see if you're correct.

Spaceship Earth

A spaceship needs an energy supply. It needs food, water, and air. It needs a good, nonpolluting waste-disposal system. It needs a stable crew and passengers. (That is, they can't fight with each other and their numbers must remain right for the size of the spaceship.) If one of these factors goes wrong, the mission fails. The spaceship could even self-destruct.

For a long time, our 4.5 billion-year-old spaceship, the earth, was traveling along just fine. All of its parts were working in synchronized harmony. Then some of the crew and passengers, the humans, started causing problems. And humans have only been aboard for a short time.

What human-caused problems does Spaceship Earth face? We are running out of energy sources and we haven't been doing enough research to develop alternative energy sources. We have food and water shortages in many parts of our spaceship. Our waste disposal methods have polluted the air, water, and soil. Our crew and passengers are fighting among themselves. But the growing number of people aboard Spaceship Earth is the biggest problem.

And the list goes on. Humans may be causing another great extinction of the earth's plants and animals. Humans have used up much of the earth's resources. We have caused the ozone hole over Antarctica. We are probably causing a global warming of our entire spaceship. Soil is losing productivity due to overuse. Forests, especially the tropical rainforests with their innumerable as yet undescribed species, are being cut and burned at a rapid rate. Even the oceans are polluted.

Take a closer look at human population growth aboard our Spaceship Earth. The following chart shows how long it has taken to add each successive billion people.

1st billion	2,000,000 years
2nd billion	130 years
3rd billion	30 years
4th billion	15 years
5th billion	12 years
6th billion	10 years (?)
7th billion	9 years (?)

Another way to look at population growth is by using a graph. The graph shows that the earth's population has grown explosively in the last 100 years. In late 1996, the earth's population reached 5.8 billion people. By 2010, the population is expected to reach 6.9 billion, and by 2020, to reach 7.6 billion.

World Population Growth Through History

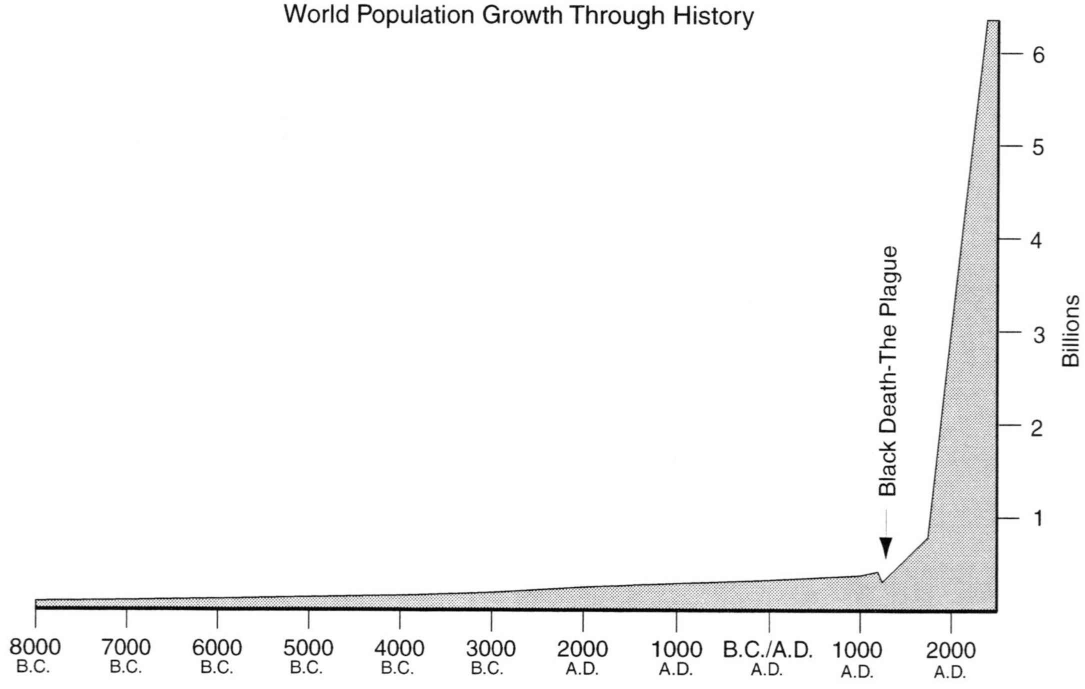

Each minute there are 180 more mouths to feed on our spaceship; each hour, 10,800; each day, 260,000; each year, 95,000,000. Imagine for a moment that all those people were born in the United States, the world's largest food producer. In three years, our population would have doubled. Could we feed all those people? Certainly. Could we feed them all in six years? Probably. In 12 years? Probably not, because our population would be about 1.4 billion and even we couldn't produce enough food.

Use reference sources such as library books, encyclopedias, world almanacs, or the Internet to find information to help you answer the following questions.

1. Compare the economies and population growth rates of the following two groups of countries. Group 1: Australia, Austria, Brazil, Canada, Czech Republic, Germany, Japan, Spain, United States; Group 2: Eritrea, Iraq, Liberia, Libya, Nigeria, Senegal, Somalia, Syria, Yemen. What conclusions can you make about the two groups?

2. List three problems facing Spaceship Earth. What solutions would you propose to help solve these problems?

GLOSSARY

aquifer	underground layer of porous sediment rock that contains water
amber	ancient tree sap that has hardened and sometimes contains fossilized insects
asthenosphere	thin zone in the earth's upper mantle that flows slowly over time and upon which the lithosphere (tectonic) plates ride
atoll	ring of low coral islands surrounding a central lagoon
barometer	instrument for measuring atmospheric pressure
barrier island	long, narrow sandy island found just off the mainland and parallel to the mainland
barrier reef	coral reef parallel to the shore of an island or a continent
black-smoker	hydrothermal vent located on a spread center on the ocean floor that spews out hot mineralized water
birth rate	number of births per 1000 people per year
chemical weathering	chemical breakdown of rock during which new products are formed in the process
cinder cone	small cone-shaped volcano that has steep sides and throws out pieces of lava and large amounts of ash
cirque	bowl-shaped basin in a mountain carved by a glacier
composite cone	large cone-shaped volcano, formed of volcanic debris and lava
continental drift	theory that the continents move over the earth's surface
continental glacier	thick ice sheet covering large parts of a continent
convection cell	space occupied by a convection current, which is the circular movement of heat in the earth's mantle
core	innermost part of the earth comprised of a liquid outer core and a solid inner core, both with a composition of iron and nickel
cross-bedding	inclined layers of sediment within beds of sediment
crust	thin outermost layer of the earth
death rate	number of deaths per 1000 people per year
desalinization	process of removing salt from seawater
drainage basin	land area that is drained by a river and its tributaries
drought	long period of dry weather
element	one of the 92 naturally occurring substances in the earth defined by the number of protons
era	portion of geologic time made up of one or more periods
erosion	the wearing away and removal of weathered rock and soil
evolution	changes in life forms through time
extinction	the wiping out of an entire species
eye	center of a hurricane where weather conditions are calm
fault	crack in the earth's crust along which movement occurs
fossil	remains or traces of organisms that have been preserved in the earth's crust

fossil fuels	resources that contain stored solar energy which can be used as fuel; coal, oil, natural gas
floodplain	land that a river covers when it overflows its banks
flood stage	point at which a river is most likely to overflow its banks
fringing reef	reef located along the shoreline of a landmass
geologic time scale	record of the earth's history based on the fossil and rock records
groundwater	water that flows or seeps downward and collects beneath the surface
half-life	measurement of the time it takes for one-half of the atoms of a radioactive element to decay
humus	decayed organic matter from plants and animals
hurricane	strong tropical storm with winds reaching at least 75 mph
hydrologic cycle	movement of water between the ground and the air by evaporation, condensation, and precipitation
hydrothermal	hot waters in the earth's crust
ice sheet	extensive body of glacial ice that is not confined to valleys
ice shelf	floating glacial ice attached to a landmass
igneous rock	rock formed from the cooling and hardening of magma
impact crater	depression made when an asteroid or comet hits the earth
index fossil	fossil that can be used to establish the relative age of the rock in which it occurs
infiltration	the soaking of water into the ground
iridium	rare element found in the clay layer at the boundary between Cretaceous and Tertiary rocks, commonly interpreted to be of extra-terrestrial origin
krill	small, shrimplike animals that are a food source for fish, birds, seals, and whales
longshore current	movement of water along the coast caused by waves striking the shore at an angle
magma	molten rock inside the earth
mantle	layer of the earth between the crust and core
metamorphic rock	rocks which have been formed from other kinds of rock by pressure and temperature
Milankovitch cycles	short-term (21,000 to 100,000 years) temperature changes caused by periodic changes in the earth's orbit, tilt, and wobble about the axis
mineral	inorganic substance that is naturally occurring and has an orderly atomic arrangement
Moho	boundary between the earth's crust and the mantle; discovered by and named after Yugoslavian seismologist Andrija Mohorovicic
moraine	ridge of unsorted sediment deposited by glacial ice
mud crack	crack in mud that occurs when the mud contracts while drying
nonrenewable resources	resources that take so much time to form that once used, they are considered gone for good
paleocurrent	ancient current whose direction of flow can be inferred by marks left in sedimentary rock
Pangaea	last supercontinent that broke up about 200 million years ago
period	division of geologic time that is smaller than an era

physical weathering processes by which rocks and minerals are broken into smaller pieces

pillow lava pillow-shaped mass of igneous rock formed by lava that was extruded underwater

plate tectonic theory idea that the earth's surface is broken into large sections or plates that move by means of convection currents beneath the plates

prehistoric overkill theory that ancient animals became extinct by hunting

P waves primary seismic waves that alternately push and pull the rock particles in the direction of wave movement

Richter scale open-ended scale describing the strength of earthquakes based on the amount of energy released

Ring of Fire area along the rim of the Pacific Ocean where approximately 75 percent of the earth's active volcanoes and earthquakes occur

rock cycle diagram that shows how rocks change from one kind to another as conditions change

saturated state of holding all the water that can be held

seafloor spreading theory that the seafloor spreads away from mid-ocean ridges as new crust is formed there

seamount flat-topped (eroded) volcano below the water surface

sediment material transported and deposited by water, wind, or ice

sedimentary rock rock formed by the compressing and cementing of sediment

seismogram chart from a seismograph that record movement of earth's crust

seismograph sensitive instrument that measures and records local earth movement (shaking) due to earthquakes

sheetwash thin layer of water that causes erosion as it flows to lower areas

shield volcano large, broad volcano formed largely of basaltic lava flows

silicon-oxygen tetrahedron basic building block of the earth's crust, comprised of a silicon atom surrounded by four oxygen atoms

spit small, sandy strip of land projecting into a body of water

spread center plate boundary formed where two plates are spreading apart

subduction downward motion of a plate as it goes into the mantle

S waves seismic waves that are slower than the primary waves; can move only through solids

tabular iceberg huge, flat icebergs that break off ice shelves

temperate climate long-term weather conditions that do not typically exhibit extremes of temperature

till unsorted glacial deposits

tornado extremely violent and destructive whirlwind extending down from a mass of dark clouds as a twisting funnel and moving over the land in a narrow path

trench narrow, elongated depression of the ocean floor, marking the place where two plates collide; subduction zone at the boundary between two converging plates

valley glacier glacier that develops in a high mountain valley and slowly flows like a river of ice down the valley